Biological Environmental Impact Studies:

THEORY AND METHODS

Biological
Environmental
Impact Studies:
THEORY AND METHODS

Diana Valiela Ward

Westwater Research Centre
University of British Columbia
Vancouver, British Columbia, Canada

Academic Press   New York   San Francisco   London   1978

A Subsidiary of Harcourt Brace Jovanovich, Publishers

ACADEMIC PRESS, INC.
111 Fifth Avenue, New York, New York 10003

United Kingdom Edition published by
ACADEMIC PRESS, INC. (LONDON) LTD.
24/28 Oval Road, London NW1 7DX

Library of Congress Cataloging in Publication Data

Ward, Diana Valiela.
    Biological environmental impact studies.

    Includes bibliographies and index.
    1. Environmental impact analysis. 2. Pollution––
Environmental aspects. I. Title.
QH545.A1W37   574.5'028   78–10595
ISBN 0–12–735350–X

Contents

Preface

Few fields of activity are as pervasive and as controversial as environmental impact analysis. Among the most useful trends apparent is the effort to integrate impact studies into the management process. Whether or not this trend will dominate future environmental impact studies, however, some form of biological investigations will continue to form part of our concern about environmental impact. The biological research presently pursued in such studies is not only costly and time-consuming but is of little value in predicting impacts, since it is based on the assumption that knowledge of the biological state of the system at present allows prediction of future impacted states. It is, of course, evident that no research approach will allow prediction of impacts with absolute certainty. However, a number of methods can be used to perform the proposed development changes on an experimental basis (e.g. on parts of the system, in comparative studies, on simulation models or physical model systems). The time and resources presently devoted to descriptive empirical efforts and guessing of impacts can be redirected to focused manipulative experimentation so that predictability of biological changes is greatly improved. In this book I have tried to specify how this type of biological environmental impact study can be approached and accomplished.

A major contribution of the book will be, I expect, to present the idea and some examples of manipulative rather than descriptive ecological studies to managers and government agents concerned with requesting and reviewing environmental impact assessment studies. A second major contribution intended is to aid the biologists performing the impact study to review rapidly a battery of approaches and methods that are applicable to manipulative impact assessment studies. Third, I hope this book will be of interest to the academic community as an aid to training biology students, many of whom will be employed to perform environmental impact studies. To accomplish these objectives biological studies are set in the environmental impact assessment context in Chapter 1; Chapters 2 through 5 are more technical and present ecological tools and guidelines for use of these tools; Chapter 6 provides some examples; Chapter 7 is a summary of the proposed approaches and a guide to the material presented in the book.

The views presented in this book are genuinely mine in the sense that many of my colleagues disagree with a number of my assertions, a phenomenon that is not surprising in a field as new and in flux as environmental impact assessment. Nevertheless, I have been greatly influenced by several research groups and individuals and draw heavily on their/our work. Among these are the research group that was directed by A. J. Forgash at Rutgers University; the research in systems analysis and resource management being done by C. S. Holling and C. J. Walters and many others in their group at the Institute of Resource Ecology, University of British Columbia; and, certainly not least, Westwater Research Centre. I wish to thank A. J. Forgash and L. M. Ward for making valuable contributions to the preparation of this book by criticizing all or part of the manuscript at a number of stages. In addition, L. M. Ward provided constant advice on statistical problems and helped with proofreading and indexing. Final responsibility for the contents, errors, and omissions of the book is, of course, mine.

Diana Valiela Ward

Biological Environmental Impact Studies:

THEORY AND METHODS

Environmental Impact Analysis

I. WHAT IS ENVIRONMENTAL IMPACT?

Environmental impact can be defined as any change in the environment that is caused by an activity or a factor. The change may be physical, chemical, biological, social, or economic; in this book I will consider only the assessment of biological change (see Section II,B). The environmental change may be caused directly by the activity, as when construction of a dam prevents salmon from swimming upriver to spawn, or may result secondarily after a series of events, as when application of an insecticide to a forest kills the insects that are food species for the forest birds and the birds subsequently move out of the affected area. The secondary impact of the insecticide factor is a reduction in forest bird populations. Accordingly, environmental impacts may be immediate or may require varying amounts of time, depending on the underlying processes, to occur. As we will see, some kinds of environmental impact are extremely subtle and complex. This book treats problems of detecting and measuring biological changes involved in environmental impacts of anthropogenic factors (human actions).

II. ENVIRONMENTAL IMPACT ASSESSMENT

A. Existing Guidelines and Previous Approaches

The last few years have seen an incredible proliferation in number and kinds of environmental impact studies performed in North

America, and we can expect this trend to continue. Such studies are performed by government agencies, academic institutions, industrial firms, and private consulting firms of many types. The studies are approached in many different ways, and the guidelines followed range from no guidelines at all to documents such as the U.S. Environmental Protection Agency's (1973) "Guidelines for Preparation of Environmental Statements," or to guidelines established by individual panels under the Canadian Federal Environmental Assessment and Review Process (Canada, 1977). The most useful available guideline document is probably SCOPE 5 (Munn, 1975). The available guidelines, however, are aimed at the general environmental impact assessment process and do not get down to comprehensive treatment of the biological studies that may be required as part of environmental impact studies. Accordingly, a variety of approaches has been used in biological environmental impact studies. One common approach has been to evaluate impact on biological systems in terms of isolated selected aspects that have apparent immediate human importance. For example, aspects of immediate economic consequence, such as effects of a factor on a commercial fishery, are often emphasized without considering other system components that may affect the fishery. Another common emphasis is the study of organisms that have shown a dramatic response to the changed factor. An example is extensive investigation of direct effects of insecticides on birds, since some populations of birds have decreased markedly as a result of insecticide use. Most other ecological components, such as invertebrates or microorganisms, that may affect the components of interest are not examined in many studies. Even in cases of impact studies that are more synecologically oriented, there is considerable variability in the parameters measured within the compartments and in whether certain important system characteristics, such as relationships among compartments, are considered. Restricted approaches to biological impact analysis often result in missing subtle but important effects and causal factors as well as effects that may only become important at some future time. Note that a variety of effects will most probably go undetected regardless of the approach used. It is clear from previous cases, however, that restricted, nonsynecological approaches are more likely to let important phenomena go undetected. Thus, it is relevant to discuss and formulate systematic but efficient approaches to examine the function of a whole ecosystem and to detect changes in that system, even if our interest is focused on selected species or components.

B. Rationale for a Biological Approach

One definition of environmental impact assessment is: "... environmental impact assessment is an activity designed to identify, predict, interpret and communicate information about the impact on man's health and well-being, of proposed human actions such as the construction of large engineering works, land reform, and legislative policy and program proposals" (Munn, 1975). The process of environmental impact assessment, however, is seen differently by different assessors, each having a viewpoint heavily influenced by, among other things, his or her disciplinary background and experience. Those of us who are biologists may admit that an overall systems approach, including economic, social, biological, and other factors, is vital to any environmental impact assessment problem. But even at this we will subconsciously weight and analyze the biological factors heavily compared to the other factors. Similarly, resource managers, statisticians, economists, sociologists, psychologists, political scientists, and others often approach environmental impact assessment problems in restricted, discipline-oriented ways. For this reason, among others, environmental impact assessments are currently often performed by interdisciplinary teams rather than by individuals or by groups of similarly trained individuals. Even using interdisciplinary teams, however, environmental impact assessments can become fragmented aggregations of pieces of information rather than unitary, comprehensive, system-oriented studies. Attempts to develop improved system-oriented environmental impact assessment procedures are presently being made (Munn, 1975; Peterman, 1975; Clark *et al.*, 1978; Holling and Clark, 1975). Holling (1978) reviewed sophisticated approaches, developed by a closely integrated working team, to embed impact assessment studies into the framework of the management process rather than to perform separate studies. Nevertheless, the separate studies will be with us for some time to come, and they are being done much more ineffectively and wastefully than is necessary.

In stating that a unitary problem-oriented approach to any environmental impact assessment is necessary, it can be argued that the present book is superfluous, since biological information should only be brought in as needed by the demands of the overall systems studies, these demands being often quite specific and restricted. More general biological studies, particularly some survey studies, tend to be (but need not be) time-consuming and expensive and often produce much information that cannot be used in the environmental impact assessment (see Chapter 6, Section I; see also Holling, 1978). The strongest point against this sort of argument is that restricted biological studies more often result in missing important impacts (see Section III). In addition, there are at least two senses in which the information contained in and the general approach taken by this book on biological aspects are very important as part of an applied systems analysis type of environmental impact assessment. First, the background (mainly qualitative) biological information needed for an environmental impact assessment may not be available, so that we cannot include relevant biological variables in the analysis. The approaches presented here are aimed at rapidly scanning a specific biological system for its essential qualitative features and measuring biological responses to the proposed changes. After gathering such general biological information, we can then use only selected aspects if and when they become relevant in the overall environmental impact analysis. These approaches may not be necessary for biological systems for which, from previously gathered data, we already have a good qualitative description (for an example, see the budworm study reviewed in Chapter 6). A second sense in which the general biological studies suggested here can be important as part of a system-oriented environmental impact assessment is as follows: In many environmental impact analyses, an assessment will be made on the basis of existing information, a course of action will be decided, and simultaneous further biological research will be indicated to study relationships that are suspected to be of interest but that could not be included in the environmental impact assessment because of lack of information. The concomitant biological research indicated can be performed with my suggested approaches in mind. Thus, even within the framework of a problem approach to environmental impact analysis, rather than a specialty-oriented approach, there is considerable need for theory and methodology within the field of ecology to study the biological function and responses of the system being analyzed.

Diagrammatically, the input of biological information into and environmental impact assessment can be pictured as in Fig. 1.1. Note

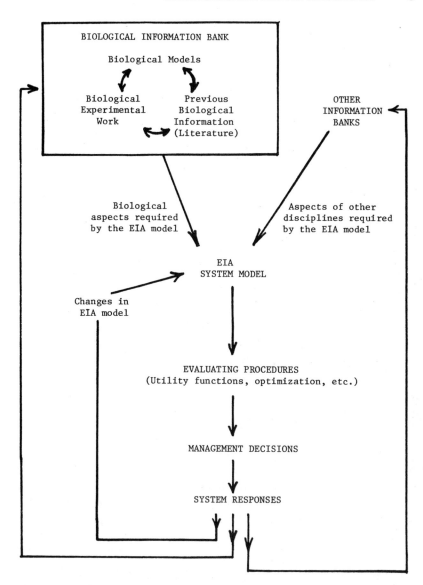

Fig. 1.1.   Diagramatic representation of the input of biological information into an environmental impact assessment (EIA). Note that the term "information bank" is used in a general sense to designate the total information available from that discipline for use by the EIA. Note direction of information flow indicated by the arrows and the various feedback loops.

that this book treats the subjects represented by the interactions between "biological models," "biological experimental work," and "previous biological information" as shown in the figure. In addition, note in the diagram that the responses shown by systems subjected to management decisions will feed information back into this biological research compartment, both in the form of new basic data and as sources of ideas for biological model modifications and new research.

In summary, the intent of this book on biological aspects of environmental impact studies is neither to be a guide for management of biological resources nor to provide a protocol for overall environmental impact assessment. It is rather to provide a battery of approaches to study biological systems such that basic functional attributes of the system can be modeled (verbally and/or mathematically), and such that important relationships between possible impacting factors and ecosystem variables can be uncovered. Of course, no possible schemes will uncover all relationships between impacting factors and biological variables. The best we can do is to do a general, systems-level study and hope to glean the important relationships and impacts. Note that within the biological context I am attempting to approach environmental impact studies with a systems orientation (see Sections III; IV,A; IV,C; IV,D).

It should also be kept in mind that my intent is not to provide a checklist or rigid protocol for biological environmental impact studies, but to present a collection of factors that may be examined and a group of effective approaches that may be used in gathering the biological information needed for biological environmental impact evaluation. Each specific study will have to be designed according to its own constraints and peculiarities (see Chapter 6). However, here I present a spectrum of indicators, tools, and concepts that may be of value in many environmental impact studies. This spectrum will hopefully offer at least some ideas, choices of indicators, and awareness of published information to each biological impact assessment problem.

III. ECOLOGY IN ENVIRONMENTAL IMPACT ASSESSMENT

Many environmental impact studies are and have been approached from only a toxicological, water quality, hydrological, etc., point of view, rather than from an ecosystem function point of view. Although in many cases a restricted approach to the detailed investigations is justifiable, there are at least two compelling reasons to precede such detailed studies by a general ecologically oriented study. First, a general ecosystem-level study will indicate which effects (toxicity, water quality, faunal changes, etc.) occur or which compartments are most affected and thus warrant further study. Second, the general study can serve to evaluate the importance of the specific effects within the context of that ecosystem's function. As examples I will cite two past cases in which the ecosystem-level study was not performed, resulting in missing subtle but important effects in the first case and in overestimating the probable importance of a specific effect in the second case.

The use of DDT in the 1940's and 1950's was accompanied by toxicological studies reporting relatively low toxicity of the compound for most vertebrate species except some species of fish (see review in Pimentel, 1971). In the 1960's, however, several studies with a more ecological orientation (Hunt, 1965; Hickey *et al.*, 1966; Woodwell *et al.*, 1967) revealed that food chain phenomena, in addition to solubility and persistence characteristics of DDT and its metabolites, brought about concentration of the compound to very high values at higher trophic levels (see Fig. 1.2). Furthermore, these DDT concentrations may continue to rise in the longer-lived organisms for years after cessation of DDT use (Harrison *et al.*, 1970). Thus, some species of birds, which in simple toxicological studies appeared not to be very susceptible to damage by DDT, were severely affected in long-term field exposures because of food-web accumulation of DDT and its metabolites to unexpectedly high levels, resulting in direct mortality and in considerable sublethal reproductive failure (Stickel and Rhodes, 1970; Risebrough *et al.*, 1970). Had an ecologically oriented study been performed earlier, the biomagnification effect could have been detected and the higher trophic level organisms pinpointed as needing study in detail before widespread use of DDT and considerable changes in populations of birds occurred. Such multiple nonobvious ecosystem effects may result not only from impacts of toxic chemicals but also of nontoxic chemicals, heat, radioactivity, creation or destruction of habitats, addition or deletion of species, induced changes in predation or parasitism, and any number of other factors.

The second important reason to begin an impact assessment investigation with an ecosystem-level study is that isolated examination of specific effects may lead to conclusions and predictions that may not be valid in the more complex context of the ecosystem. For example,

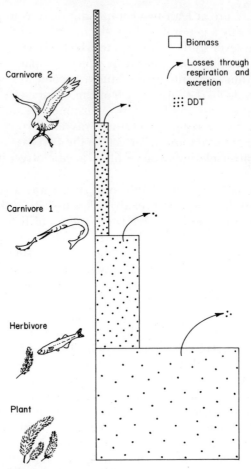

Fig. 1.2. Schematic diagram showing concentration of DDT residues as they are passed along a simple food chain. As biomass is transferred from one link of the food chain to another, usually more than one-half of it is consumed in respiration or is excreted (note decrease in size of biomass boxes); the remainder forms new biomass. In contrast, the losses of DDT residues (shown by arrows) along the food chain are small in proportion to the amount that is transferred to higher links. The net effect is high DDT residue concentrations in the carnivores. (From Woodwell, 1967.)

Wurster (1968) reported that DDT at about 0.01 ppm inhibited photosynthesis in four species of marine phytoplankton and in a natural mixed phytoplankton culture when tested in flasks in the laboratory. However, a number of later studies indicated that extrapolation of these laboratory results to predict a reduction in algal photosynthesis under field conditions is not justified. First, several studies showed that different algal species were markedly differentially sensitive to photosynthetic inhibition by DDT (Menzel *et al.*, 1970). Thus, in many mixed communities overall photosynthesis may not be at all affected if the less sensitive algal species photosynthesize at increased rates, perhaps favored by less competition for nutrients from the more DDT-sensitive species. We can still expect some considerable ecosystem effects because of the DDT, since the algal species composition would be affected by the differential sensitivity of various species. In turn, the change in algal species composition may lead to changes in the species composition of the zooplankton grazers and so on up the food web. This ecosystem effect, however, is quite different qualitatively from an expected reduction in photosynthetic rate by an algal community. In fact, under the conditions just described, the overall photosynthetic rate may remain unaffected by the presence of DDT. Second, various studies have shown that other factors present in natural ecosystems and absent from the laboratory culture flasks may drastically reduce or totally eliminate the ability of DDT to inhibit algal photosynthesis. When sediments are added to the culture flasks, for example, the inhibitory effect of DDT is significantly reduced (Gillott *et al.*, 1975). The factor accounting for these results is adsorption of DDT on the sediment particles. Since DDT is hydrophobic but quite soluble in fats, in a natural aquatic ecosystem the DDT would also be taken up by zooplankton, organic chemicals in debris and sediments, and a number of other animals and plants. Thus, the amount of DDT available to affect the algal cells would be severely limited and may not cause significant photosynthetic inhibition. Although Wurster (1968) suggests that concentrations of DDT in natural waters, although quite low, "... are likely to be more constant, the DDT being replaced from persistent residues in mud, detritus, runoff water, and other sources as it is absorbed by cells ..." this replacement seems unlikely to be significant in view of the very high affinity of natural sediments for DDT. Oloffs *et al.* (1973) incubated five chlorinated hydrocarbons, including DDT, in natural waters with sediments and found that after

6 weeks all detectable amounts of pesticides (except for lindane) had moved into the sediments. Clearly, an *in situ* ecosystem study measuring the DDT (and DDT break-down product) levels actually found in algal cells and in other system compartments and measuring the photosynthetic inhibition resulting from those same levels in algae is needed to evaluate this relationship.

From these two examples of DDT ecosystem dynamics, which produced unexpected results, it is clear that a limited nonecological approach to impact studies can lead to both underestimation (as in the case of biological magnification of DDT) and overestimation (as in the case of algal photosynthetic inhibition by DDT) of the impact of a factor on a biological system.

IV. GENERAL CONCEPTS FOR BIOLOGICAL IMPACT STUDIES

Four general concepts for biological impact studies are presented here. These principles arise from application of theoretical ecological concepts to impact evaluation problems and from consideration of various difficulties in many past impact evaluation studies. As general concepts they apply to most situations, regardless of the factor causing the biological impact or of the particular ecosystem being examined.

A. Synergy

Fuller (1969) defines synergy as the "... behavior of whole systems unpredicted by the separately observed behaviors of any of the system's separate parts or any subassembly of the system's parts." This concept is of great importance in ecology, as has been shown by the above example of biological magnification of DDT. In ecosystems, study of the gross or general features is essential for predictability of behavior. Reductionist approaches elucidating underlying mechanisms without separate study of overall system characteristics can be severely misleading and incomplete (Mann, 1975). Kerr (1974) elaborates on a similar point, differentiating "macrosystem" from "microsystem" properties as well as "emergent" properties.

Several techniques are available to aid in a synergistic approach to biological impact studies. First, *in situ* studies should be performed on the natural system to discover overall system characteristics. This is in contrast to one traditional approach, consisting of removing component organisms and testing them in laboratory situations. In these *in situ* studies it is important to uncover system characteristics (e.g., productivity and energy flow patterns, species interactions) rather than characteristics of system components (e.g., population density of a species, temperature tolerance of a species, etc.). Second, modeling techniques, both mathematical and physical, are available to explore system characteristics experimentally. Using models in conjunction with *in situ* experimental studies, the real system's overall function and its components' interactions can be simulated and studied.

B. Experimental Control and Causality: A Multifaceted Approach

Establishing satisfactory control conditions and cause and effect relationships is a problem of importance in all experimental studies, including ecological experiments. In ecology, however, it is often even more difficult than in other areas to eliminate some sources of environmental variability temporarily so that selected manipulated variables can be tested as causal factors.

A number of techniques can be used to control and/or monitor environmental variability in a field study. Control can be enhanced by (1) counterbalancing known sources of variability, (2) side-by-side comparisons, and (3) before and after comparisons. Monitoring and examining methods for environmental variability include (a) measurement of all variables that appear possibly relevant and statistical analysis to check for significant correlations and (b) mathematical modeling and simulation runs. Note that this is only one of the many uses of mathematical modeling, a topic discussed in Chapter 3.

Counterbalancing known sources of variability can be very useful in distributing known biases among all experimental and control conditions (see Fig. 1.3a). For example, suppose we want to sample the grass standing crop in a field under experimental treatment and in a control condition. If we know that there is a moisture gradient in the field, with wetter conditions as one approaches a creek at one end, we will probably want to counterbalance for this source of variability by having an equal distribution of samples near the creek and far from the creek in each condition of the experiment. However, if we see no obvious gradients in nonmanipulated environmental variables, we will probably want to randomize the sample placements. As another example, in a study where two different observers are to collect the data, experimenter variability can be counterbalanced by designing either a random or a systematic sampling schedule, in both space and time, which distributes the two experimenters equally among experimental conditions. In a random schedule, which samples are to be taken by which experimenter can be determined by use of a random numbers table. In a systematic scheme, each of the two experimenters can be directed to take, for example, every other sample (spatially) or perhaps the morning sample on one day and the afternoon sample on the following day, and so on, thus counterbalancing for the suspected variability between experimenters. For each case, spatial, time, and situational conditions should be considered when designing a counterbalancing scheme. Note that here I am discussing only some general aspects of sampling that have been particularly relevant to impact studies and that are not discussed elsewhere in the impact literature. Several further sampling questions are discussed in Chapter 4. For a more comprehensive coverage of sampling methods and theory see Cochran (1963), Poole (1974), Southwood (1966), and Yates (1963). Control over an experimental situation usually implies comparison of an experimentally manipulated condition with an untreated condition. In side-by-side comparisons (see Fig. 1.3b), the two conditions are monitored starting at the same time and are usually as similar and as spatially close to each other as possible, since spatial separation usually entails climatic and geographic differences. However, very close proximity of the treated and untreated conditions is sometimes impossible because of problems of contamination of the control condition by the treated condition, of finding two large enough suitable sites, etc. Where these problems become extreme, it is sometimes necessary to abandon side-by-side comparisons entirely and establish before and after comparisons (see Fig. 1.3b). For example, in a shallow bay area where the heated effluent from a nuclear power plant is to be dissipated, there may not be any nearby unaffected comparable area to use as a control sampling site. If so, we can do as much sampling as possible in the area before the plant becomes operational and the heated effluent affects the biota. These preoperational data can then be compared to those collected at the same site after the thermal influence is established. This approach has severe limitations, since the lapse in time between the two sampling programs will probably effect a number of other changes, such as biogenic population fluctuations, seasonal and yearly climate changes, etc., in addition to

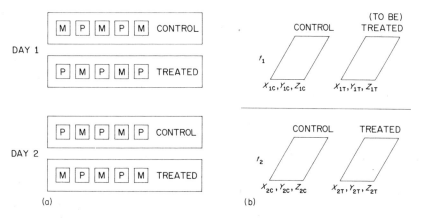

Fig. 1.3. Methods for controlling variability in a field experiment. (a) Counterbalancing known sources of variability. In this example, counterbalancing for differences between experimenters "M" and "P" is diagramed. On day 1, "M" samples three control sites and two treated sites, while "P" samples two control sites and three treated sites; on day 2, the reverse is done. On day 1, "M" takes the first sample on the control plot while "P" takes the first sample on the treated plot; on day 3 the reverse is done. The pattern is repeated on succeeding days. Simultaneous counterbalancing for other sources of variability can be designed along similar lines. (b) Experimental control using side-by-side comparisons and before and after comparisons. Side-by-side comparisons: Two matched plots or conditions are selected at the beginning of the experiment. To match plots, two sites are chosen which are as similar as possible on visual inspection. Then a number of environmental, physiographic, and biological variables are measured with appropriate replication of sampling. If there are no significant differences in these variables between control and experimental plots, the two can be considered matched. At time t_1, variable X, Y, and Z are measured in the control condition (X_{1C}, Y_{1C}, Z_{1C}) and in the treated condition (X_{1T}, Y_{1T}, Z_{1T}). At a later time t_2 (and, if appropriate, at t_3, t_4, t_5, etc.), the same measurements are made again and the values X_{2C}, Y_{2C}, Z_{2C} and X_{2T}, Y_{2T}, Z_{2T} obtained. Then statistical tests are done to compare the changes over time in the variables measured in the control plot with those measured in the treated plot. If these comparisons yield significant differences, the treatment, rather than some other factor, can be argued to have caused the observed changes in the treated plot not occurring in the control plot. Note that there will be changes in the control condition from t_1 to t_2; the same changes should also occur in the treated condition, but this latter condition will show additional changes due to the treatment. Methods for partialling out the various sources of variability are described in statistics textbooks (e.g., Sokal and Rohlf, 1969). Before and after comparisons: In these situations, only the plot or condition to be treated is available (treated at t_1 and t_2 in the figure). The appropriate variables are measured at t_1 and t_2 and changes over time are examined for the hypothesized treatment effects. Since any observed changes can also be due to other, uncontrolled, factors, such changes must be statistically treated (by methods such as multiple regression or analysis of covariance) to separate the effects of the treatment from those of the measured, uncontrolled factors.

the effects of the heat addition. However, it can still be useful if used in conjunction with a number of other techniques, such as those mentioned here, which allow discrimination of some of these other variable effects from the effects of the heat addition. Of course, the combination of both a side-by-side comparison and a before and after comparison is most desirable as a control design.

When environmental variability cannot be controlled by effective counterbalancing methods and side-by-side control comparisons, we can monitor the environmental variables and use statistical techniques for evaluating the probability that these variables are related in some way to the main effect we are measuring. The relationships are usually examined by regression and correlation techniques, explained in statistics textbooks such as Sokal and Rohlf (1969) or Snedecor and Cochran (1967), or in the useful sections on regression and correlation in Poole (1974). If we find significant relationships between a monitored environmental variable and the main effect measured, then the relationship between the manipulated variable and the main effect (impact) may not be as simple as we think and will need further investigation, including the role of the monitored environmental variable. On the other hand, finding no significant relationship between monitored environmental variables and the main effect measured, but a significant relationship between the manipulated variable and the main effect measured, increases our confidence in the established relationship between the latter two factors.

A further way to examine the validity of a hypothetical cause and effect relationship, and the effects of environmental variability on this relationship, is mathematical modeling and computer simulation. Once we have established a tentative relationship between a manipulated variable and the main effect measured, we can construct a model of this relationship, either as a set of equations or a series of FORTRAN or other computer language statements (see Chapter 3 for a fuller description of this process). Then we can test, at least to some degree, the validity of this cause and effect relationship by picking new values for the manipulated variable, calculating the new values for the main effect using the model, and experimentally checking the calculated (predicted) values. If these values agree for a number of cases, we begin to gain some confidence in our hypothetical cause and effect relationship; if there is little or no agreement, we are forced to reexamine our hypotheses and to guess at the relationship again, using the insight gained from the kind of lack of agreement between predicted and measured values. Finally, we can further examine our hypothetical relationship by varying environmental factors in our model, using simulation runs, to delimit the conditions under which the cause and effect relationship would hold. In a computer simulation we can allow for setting different values for those environmental variables that we cannot manipulate in the field. We can vary these values within reasonable ranges, as determined by long-term field monitoring data. If the predictions of the model, after simulation runs, are still stable or match observed values of the main effect measured, we are again supported in our hypothetical relationships. However, if varying values of environmental parameters within reason produces breakdown of predictability of the main effect as a function of the manipulated variable, we are again forced to reexamine that function. Thus, there are several ways in which modeling and computer simulation can be used to examine and improve hypotheses about cause and effect relationships between the manipulated variable and the dependent impact variables.

In sum, we have a number of techniques to either control or monitor and examine variability in an ecological impact experiment. The ideal situation is to use all of these methods simultaneously. The results may all support the initial hypotheses about the impact effects and their causal factors. When these methods do not converge on the same relationships, the conflicting results may aid us to construct revised or new hypotheses, which can again be examined with the same multitechnique approach. When it is impossible to use all or most of these methods, it is important to try to use at least a few. Reliance on a single method of controlling or monitoring experimental variability in a natural ecosystem may lead to incorrect acceptance or rejection of a cause and effect hypothesis. Note that these methods all pertain to field experimental data. Additional sources of agreement or disagreement may also be obtained by doing separate laboratory experiments about the hypothesis in question.

C. On Measures of Ecosystem Change

We have seen in previous sections that to evaluate environmental impact we need to measure changes in ecosystem level processes. A number of measures are available to describe the function of ecosystems, both qualitatively and quantitatively. In the past, some use has been made of one or a few selected ecological measures, such as species diversity indices, to evaluate environmental impact. However, it is dangerous to rely on one or a few measures, since changes in one measure are not necessarily coincident with changes in the overall ecosystem. For example, Teal and Valiela (1973) found no changes in species composition on a Massachusetts salt marsh after sewage sludge treatments. However, the sludge treatments effected a considerable change in the primary productivity of the marsh. Thus, measuring only species composition would have given the erroneous impression that the ecosystem was not significantly affected. Note that species composition studies and species diversity indices can be useful, as discussed in Chapter 2. However, the use of any one or only a few measures to assess ecological change is not advisable. What is needed is a qualitative description of the ecosystem in terms of structure and function, and a range of quantitative measures to support the qualitative picture. The set of measures to be used for any particular impact assessment study will depend on the nature of the ecosystem in question and of the impact in question. In Chapter 2 I suggest a broad set of ecological measures that may be of use in selecting specific sets of measures. It is important to select a variety of measures of change for each study attempting to assess ecological change.

D. Interrelationship of Structure, Function, and Time

In constructing the qualitative description of the ecosystem needed to assess changes in that system, we need to pay attention to both structural and functional dimensions. Examples of structural dimensions of ecosystems are density of species, nature of feeding relationships, and species diversity. Examples of functional dimensions are productivity, nature of change in population densities with predation, and energy flow.

Excessive reliance on measuring either structural or functional characteristics may result in serious omissions from the overall description of the ecosystem and its changes. For example, we may describe species composition quite accurately and see considerable changes in this structural characteristic due to an anthropogenic factor. However, this change may be of little or no consequence functionally if the new species replacing the old perform a similar functional role in the ecosystem in terms of productivity, species interactions, etc. (as suggested by Walters and Efford, 1972; Walters *et al.*, 1978). Similarly, changes in functional parameters, for example, productivity, by themselves will not indicate whether the system has been altered by replacement or disappearance of species, changes in trophic relationships, etc., or whether it remains structurally unchanged. Thus, it is important to include both functional and structural measures in our set of indicators of ecosystem change.

Time is, of course, also an important factor, and is related to structural and functional ecological characteristics. The time course of ecosystem alterations will vary widely from instantaneous response (e.g., instantaneous kill of a species by a toxicant) to very long-term effects, some of which may be undetectable for long periods of time [e.g., the DDT accumulation effects depending on the life span of species, as proposed by Harrison *et al.* (1970)]. Thus, it is important to look for changes occurring at widely differing rates in an impacted ecosystem. This is another compelling reason to use as many different approaches as possible to assess ecological change. That is, we may get advance indications that there may be long-term effects by using laboratory studies, mathematical modeling, and physical model systems, rather than only detecting such changes later, in long-term field studies. For example, in the case of the effects of DDT on birds, if a spectrum of laboratory tests guided by modeling and/or interactive field studies (see Chapter 5) had been performed on the effects of DDT and its break-down products on bird physiology when DDT use began, we would have detected the reproductive system effects when certain DDT/DDE levels were reached. Thus, we would have had an indication of the population changes in birds of prey that we saw much later and whose basis we only learned *post hoc*. Therefore, it is of value to study many possible effects using as many different approaches as possible, since such studies may help to overcome the difficulty inherent in detecting changes occurring in different, and especially in long, time frames.

V. APPROACHES TO BIOLOGICAL IMPACT STUDIES

From the four general concepts for impact studies presented above, we can conclude that to begin to analyze environmental impact on an ecosystem critically, we must study the entire system in its gross characteristics as well as its parts. We must use a number of different control and experimental comparison techniques, rather than one or two, to be confident of cause and effect relationships between the anthropogenic changes and their supposed impacts. We need to use a large set of measures, the set varying with the situation and including qualitative and quantitative features, to characterize ecosystem change, rather than expect one or a few indices to indicate this change. Further, this set of measures should include both structural and functional characteristics and their interrelationships. In addition, we must proceed with awareness of, and use techniques to detect, changes in structure and function occurring in widely differing time frames.

How can we achieve these ambitious goals? As impossible as this task may seem, we can at least attempt to satisfy these requirements by using a multifaceted approach including the following plan and categories of studies. First, do a rapid preliminary literature/field survey of the system to obtain an initial description. On this basis, construct a qualitative explicit model of the major factors controlling the biological components in question or of the major ecological processes occurring in the ecosystem in question. Based on this analysis, and influenced by the terms of reference of the impact study (see Chapter 2, Section V and Chapters 3, 6, and 7), list predicted biological effects of the development factor(s) on the biological components or the ecosystem processes in question. Design formal hypothesis-testing field and/or laboratory experiments to test the predicted effects of the development factor(s) in some experimental system (see Chapter 4). Based on the results, modify the previous explicit model and on that basis predict impacts of the development factor(s) on the species or ecosystem in question.

In the chapters that follow more specific rationale and methodology will be provided to support the value of this approach. Chapters 2, 3, 4, and 5 deal with the tools and background needed to pursue this approach. Chapters 6 and 7 deal with approaches and strategies. Chapter 6 provides some examples of studies performed according to these approaches even within very limited time and resource conditions. Chapter 7 includes a restatement and elaboration of the desirable sequence of general steps described above, and a comparison of this sequence with that followed in many existing biological impact studies.

REFERENCES

Canada (1977). "A Guide to the Federal Environmental Assessment and Review Process." Dep. Fish. Environ., Ottawa.

Clark, W. C., Jones, D. D., and Holling, C. S. (1978). *Ecol. Model.* In press.

Cochran, W. G. (1963). "Sampling Techniques," 2nd Ed. Wiley, New York.

Fuller, R. B. (1969). "Operating Manual for Spaceship Earth." Southern Illinois Univ. Press, Carbondale, Illinois.

Gillott, M. A., Floyd, G. L., and Ward, D. V. (1975). *Environ. Entomol.* **4**, 621–624.

Harrison, H. L., Loucks, O. L., Mitchell, J. W., Parkhurst, D. F., Tracy, C. R., Watts, D. G., and Yannacone, V. J., Jr. (1970). *Science* **170**, 503–508.

Hickey, J. J., Keith, J. A., and Coon, F. B. (1966). *J. Appl. Ecol.* **3**, Suppl., 141–154.

Holling, C. S., ed. (1978). "Adaptive Environmental Assessment and Management." Wiley, New York.

Holling, C. S., and Clark, W. C. (1975). *In* "Unifying Concepts in Ecology" (W. H. Van Dobben and R. H. Lowe-McConnell, eds.), pp. 247–251. Junk, The Hague.

Hunt, L. B. (1965). *U.S. Fish. Wildlf. Serv., Circ.* No. 226, 12–13.

Kerr, S. R. (1974). *Proc. Int. Congr. Ecol., 1st, The Hague* pp. 69–74. Cent. Agric. Publ., Doc., Wageningen.

Mann, K. M. (1975). *In* "Estuarine Research" (L. E. Cronin, ed.), Vol. 1, pp. 634–644. Academic Press, New York.

Menzel, D. W., Anderson, J., and Randtke, A. (1970). *Science* **167**, 1724–1726.

Munn, R. E., ed. (1975). "Environmental Impact Assessment: Principles and Procedures," SCOPE (Scientific Committee on Problems of the Environment), Rep. No. 5. Int. Counc. Sci. Unions, Toronto.

Oloffs, P. C., Albright, L. J., Szeto, S. Y., and Lau, J. (1973). *J. Fish. Res. Board Can.* **30**, 1619–1623.

Peterman, R. M. (1975). *J. Fish. Res. Board Can.* **32**, 2179–2188.

Pimentel, D. (1971). "Ecological Effects of Pesticides on Non-Target Species." Exec. Off. Pres., Off. Sci. Technol., Washington, D.C.

Poole, R. W. (1974). "An Introduction to Quantitative Ecology." McGraw-Hill, New York.

Risebrough, R. W., Davis, J., and Anderson, D. W. (1970). *In* "The Biological Impact of Pesticides in the Environment," (J. W. Gillett, ed.) Environmental Health Sci. Ser., No. 1, pp. 40–53. Oregon State Univ., Corvallis.

Snedecor, G. W., and Cochran, W. G. (1967). "Statistical Methods," 6th Ed. Iowa State Univ. Press, Ames.

Sokal, R. R., and Rohlf, F. J. (1969). "Biometry: The Principles and Practice of Statistics in Biological Research." Freeman, San Francisco, California.

Southwood, T. R. E. (1966). "Ecological Methods, with Particular Reference to the Study of Insect Populations." Methuen, London.

Stickel, L. F., and Rhodes, L. I. (1970). *In* "The Biological Impact of Pesticides in the Environment," (J. W. Gillett, ed.) Environmental Health Sci. Ser., No. 1, pp. 31–35. Oregon State Univ., Corvallis.

Teal, J. M., and Valiela, I. (1973). *Oceanus* **17**, 7–10.

U.S. Environmental Protection Agency (1973). "Guidelines for Preparation of Environmental Statements." U.S. Gov. Print. Off., Washington, D.C.

Walters, C. J., and Efford, I. E. (1972). *Oecologia* **11**, 33–44.
Walters, C. J., Park, R., and Koonce, J. (1978). *In* "Synthesis of IBP Freshwater Results" (E. D. LeCren, ed.), Chapter 10. Blackwell, Oxford.
Woodwell, G. M. (1967). *Sc. Am.* **216**, 24–31.
Woodwell, G. M., Wurster, C. F., and Isaacson, P. A. (1967). *Science* **156**, 821–824.
Wurster, C. F., Jr. (1968). *Science* **159**, 1474–1475.
Yates, F. (1963). "Sampling Methods for Censuses and Surveys," 3rd Ed. Griffin, London.

2

The Field Survey: Preliminary System Analysis

I. FIELD SURVEY OBJECTIVES

Before the impact of a specific human influence on an environment can be assessed, it is obviously necessary to have an initial description of that environment, i.e., to investigate the general structure and function of the system to be affected. This preliminary investigation is often overlooked or performed inappropriately, resulting in severe impairment of the ability to detect changes due to the experimental variable. It should be emphasized that I do not mean, by an initial field survey, the typical lengthy cataloging and checklisting of the existing species. A field survey, to fulfill its purpose of suggesting hypotheses for later testing, should uncover possible functional relationships among important variables in the ecosystem studied and possible functional relationships among those variables and the proposed impacting factors. To obtain this type of information, a broad range of ecological traits must be examined. A basic knowledge of the main structure and processes involved, whether it be derived from the literature, from *in situ* measurements, or from both, is essential before critical experiments can be conducted.

There are two main reasons for the importance of the field survey. First, it is very difficult to choose the important variables to be tested for possible impact without knowing the peculiarities of the system in question. For example, we may decide on theoretical grounds that one of the major variables to test for impact by a pollutant is the energy processing by herbivores. If this decision were made about a mangrove swamp ecosystem, the study would probably be wasteful of time and effort and not yield the most valuable information, since mangrove swamps are primarily detritus-based food webs, routing little of the energy processing through herbivores (Heald, 1969; Odum, 1970). It would be of much higher priority to measure effects on primary productivity and detritivore processes rather than on herbivory. In this case, the field survey information is available in the literature; in other cases, experimental work may be necessary to gather this type of information. Thus the investment of time and effort in the prestudy can be a great saving in the long run and is insurance against gathering extensive nonessential data at the expense of the essential and against missing important system characteristics. If one is fortunate, much published or private data will exist on the ecosystem in question, and minimal study will be needed to ensure understanding of major characteristics. At the other extreme, little information will be available and an overall systems study will be needed before manipulative work involving the proposed impacting factor can begin.

The second reason for the importance of the field survey is that the prestudy can be a form of experimental control (the "before and after comparison" described in Chapter 1), providing a set of data that can be compared with those of later studies on the impacted system. Clearly, this control condition does not adequately substitute for the "side-by-side comparisons" described in Chapter 1, but is nevertheless valuable as a time comparison reference point. For some cases, it is the only control condition possible, for example, if an installation or pollutant will impinge on an environment in such a way that no unaffected comparable control areas will be available for simultaneous study. In such cases, it is desirable to have information about the environment before the installation or pollutant addition begins so that some comparisons can be made with the disturbed environment at a later time.

There are strong arguments, therefore, for a good initial system description in the form of a field/literature survey. In the sections that follow I will suggest and discuss some major ecological characteristics that can be investigated in a field survey and, later, in experimental and comparative impact studies. Some, most, or all of these may be applicable to any single study. After the field-literature survey, these characteristics can be evaluated as variables to be manipulated and/or studied in the overall research program.

Although there will be some overlap in the following classification, for convenience I have divided characteristics of ecosystems into two groups: structural variables, such as species composition and feeding relationships, and functional variables, such as productivity and energy flow. As we have seen in Chapter 1, both types of characteristics should be included in an impact study. The basic questions involved are: What is the system like? Of what is it composed, and how does it operate?

II. SYSTEM STRUCTURAL STUDY

A. Species Composition and Abundance

Part of the system description involves study of the major species present and of their abundance. In many cases, identification down to the species level will be needed *only* for the most common organisms, with the rarer species being determined at most as to the number of species present and major groups to which they belong. This procedure is almost always necessary and desirable to avoid a total impasse at the taxonomic level of the study. Population densities may be measured as crude densities, ecological densities, or relative abundance indices, depending on the nature of the ecosystem in question. Crude densities are simply the number or biomass of organisms per unit of total area or volume. For example, we may find that in an oak–hickory forest under study the crude density of black oak is 200 trees per hectare. This measure is obtained by sampling at various places in the forest, without regard to whether the site is a typical forest site or a lake or river area. Crude density measures the number of organisms per unit of space. If, on the other hand, we wish to know the population density of black oak in those areas where the species would or could normally occur, we would measure the numbers or biomass of black oak trees per unit area only in those areas. Thus we would exclude lakes and river beds from our survey, and the figure in numbers of black oak trees per hectare (of occupiable space) would be a somewhat larger number, expressing ecological density.

In many cases where environmental impact is concerned, a third measure of population density may be not only appropriate but also economical of time and effort. This is a relative abundance index, which may be manufactured for specific cases according to convenience and method of survey. Some examples of relative abundance indices are the number of ducks harvested per man-day of hunting effort; the percent of *Calanus* in each sample of plankton (or the "abundance" of *Calanus*); or in the number of woodpeckers seen per transect walked in a bird census. Note that these do not relate the number of organisms to a unit of area or volume, but are relative to some sampling procedure unit. Relative abundance indices of this nature may save a lot of painstaking measurement involved in absolute density estimates and yet give numbers that can be used to compare experimental versus control conditions as well as to detect any increases or decreases in numbers or biomass over time.

The methodology of estimating population densities is very varied and depends on the kind of organism and habitat in question. Most often, there are several methods available which must be evaluated for use. In some cases, several methods will be adopted to provide comparative data. Before attempting field determinations of population densities, specialized reference works on the methodology for each kind of organism of interest should be studied. Southwood (1966) is one useful reference for this purpose. The methods adopted for various types of organisms by several research groups are presented as illustrations in Chapter 6.

It should be noted that changes in population densities are to be expected in undisturbed systems, since there seem to be multiple stable points in most systems (Holling, 1973; Neave, 1953; Larkin, 1971; Sutherland, 1974). Normal fluctuations in the populations must not be ascribed to impact variables. In this connection, side-by-side comparisons (see Chapter 1) are invaluable. Other sources of information on population fluctuations, such as published or private data on previous years, may be sought as well. It is important to try to differentiate a temporary reversible population oscillation from a permanent one leading to a new stable point and from a destabilizing one leading to disappearance. An example of a relatively simple study examining population changes and making hypotheses about qualitative changes in the system is Breen and Mann's (1976) study of lobsters, kelp, and sea urchins. Holling (1973) should be read for a detailed discussion of stability problems and of quantitative versus qualitative changes in natural systems. Throughout environmental analysis problems, it is important to try to detect effects (such as species disappearance) that will bring about qualitative changes in the system. We are examining the system to see if it will persist as well as to see how its various variables will fluctuate. We examine the fluctuations to see if they are likely to lead to qualitative and/or permanent changes in basic features. Holling (1973) suggests that, contrary to common expectation, ". . . instability, in the sense of large fluctuations, may introduce a resilience and a capacity to persist."

B. Feeding Relationships

Early in the study of what organisms are present in the system, it is important to begin to make as many observations as possible on the feeding relationships of at least the major species present. Observations of feeding activity can be supplemented with inferences from structural features of organisms, such as type of feeding apparatus present. Of course, information from the literature may be available as well. In some cases, when little food web information is available and such information becomes important, it may be necessary to perform separate studies on the food sources of various species. The two most common experimental approaches to this type of study are gut content analyses (Neal et al., 1973), particularly for vertebrates, and radio-isotope labeling experiments (Odum and Kuenzler, 1963; Marples, 1966; Shure, 1970). The gut content methods has several important drawbacks. Fluids and soft tissues may not be detected, even though they may constitute the major source of food. A large collection of each species is needed to get a representative indication of diet through seasonal changes, various locations, and individual variability. Commonly, taxonomic recognition of gut contents is difficult or impossible, since the specimens are most often fragments of organisms. Other techniques recently developed to trace food web connections are based on antiserum reactions (Dempster, 1969; Reynoldson and Young, 1963; Young et al., 1964) and on fatty acid analyses (Jeffries, 1972).

As usual, several sources of information are better than one, although each case must be evaluated individually for the extent of experimentation needed to acquire the desired level of certainty in food web information. In some cases, such as when modeling the effect of a predator on a prey species population in detail, knowing the major food sources of the predator under various environmental conditions and at different times may become of major concern. In other cases, only general food web connection information is needed as part of the structural preimpact study in order to make comparisons with the impacted state of the system.

C. Ecological Dominance and Key Species

An important attribute of an ecosystem is the pattern of importance of its various species. It is of value to determine if an impact involves changes in the species dominance patterns, since changes in such patterns sometimes entail a total shift, both qualitatively and quantitatively, in an ecosystem.

The terms ecological dominance and key species have both been used extensively to refer to the importance of a species in maintaining the particular structure and overall function of the community of which the species is a part. The two terms have been used with somewhat different meanings, however. Ecological dominance usually refers to species that control a major portion of the community energy flow (see Sections III,A and III,B). Usually, species controlling a major portion of community energy flow have high relative abundances, high biomass, or high productivity; thus, these parameters are usually measured to determine the dominant species. Whittaker (1965) points out that the best indication of dominance is probably the productivity of a species, since the productivity also reflects the species' biological activity and the share of environmental resources it utilizes. Odum (1957) determined the standing crop and energy flow control of the various organisms in the Silver Springs, Florida, community. The eelgrass, *Sagittaria lorata*, had the greatest biomass and was the dominant producer in the system. Although often ecological dominants are also visual dominants (those species we can see as being very abundant in an environment), this is not always the case. Small organisms with a high rate of population turnover, such as some planktonic species, can control a major portion of the community energy flow but not appear to be dominant in that the observer may only see a small standing crop at any one time.

In addition to identification of ecological dominants, dominance analysis can be expanded to compute indices of dominance, which are measures of the degree of concentration of dominance. That is, dominance indices are high when just a few species out of many in a community are dominant, and the indices are low when dominance is shared more extensively among species. Whittaker (1965) reviews different indices of dominance that may be used for computation. Changes in these indices can be of value as an indication of structural and perhaps of functional changes in a community in response to perturbations.

The term key species (also called keystone species, foundation species, etc., by various authors) is in some ways similar to the concept of ecological dominance. It has been used to refer to a species that has a strong influence on most other organisms in the community and, if removed, would drastically change the community characteristics (Paine, 1969; Dayton, 1971). Such strong influence is usually demonstrated by experiments in which various species are removed from a community and the resulting effects are studied. For example, Dayton (1975) found that the alga *Hedophyllum sessile* was a keystone species in areas of moderate wave exposure under study, since removal of various algae showed that *Hedophyllum* displaced many species and protected other (understory) species. Unfortunately, the term ecological dominance has also been used to refer to this definition of key species (Dayton, 1975). Ecological dominants in the energy control sense and key species as defined here often do not coincide in a community (Dayton, 1975; Paine, 1974).

Whether ecological dominants, or keystone species, or both, are determined, the dominant species in a community are usually few in number [although exceptions can be expected in stable, mature, highly diverse, and complex communities (see Odum, 1971)] and may be detected empirically. Initially, dominant species may be determined loosely by visual indications and literature review; if necessary, measurements of productivity, abundance, or biomass and removal experiments can be performed to identify dominant species. It becomes important to identify the dominant species in the field survey so that experimental measurement and manipulation can focus on those species that, if affected by the proposed environmental change, will induce important changes in the overall system.

D. Indicator Species and Ecological Indicators

Often in impact evaluation studies we seek methods of minimizing the number of variables we must measure to indicate the condition or nature of the environment we are studying. Indicator species have been used to evaluate the prevailing conditions in a variety of situations. For example, Eliassen (1952) summarizes the changes in the species of fish, larger invertebrates, and plankton that are found in a stream as it changes from the unpolluted state to one of untreated sewage pollution and back to the unpolluted condition (see Fig. 2.1). Thus, finding a particular set of species in a stream can indicate whether a stream is polluted or not. This determination can save a lot of the time and work that would be needed to chemically analyze the stream conditions. The use of indicator species has important limitations which should be kept in mind. First, it is rare to find a single species or restricted group that can serve as a valid ecological indicator (for one example, see Stockner and Benson, 1967). Very few species are narrowly restricted to a specific set of environmental conditions, and even then such species are not usually the most numerous or dominant ones in the community. More often, a set of species, rather than a single species, must be used to indicate environmental conditions (see Eliassen, 1952). Even better, numerical relationships between species, populations, and communities, can be used as indicators; examples are characteristics discussed throughout this chapter. A second limitation is that small species with rapid turnover rates usually are not suitable as indicator species, since they are not often stable in their presence in an environment. Various species of algae have been sought as ecological indicators (Rawson, 1956; Patrick,

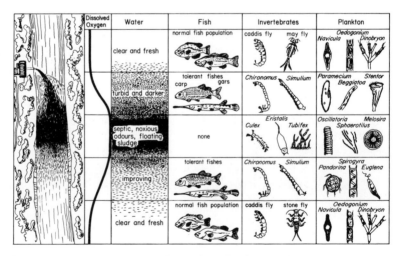

Fig. 2.1. An example of assemblages of species that can be used as indicators of pollution. (From Eliassen, 1952.)

1965) but have not been generally satisfactory (Wilber, 1969). Larger organisms are usually more stable when they are present in an environment; their generation time is longer, and their turnover rate is smaller. A third limitation is that it is important to consider the amount of work and time that it takes to establish a species or a set of species as reliable indicators of a certain set of conditions, and to compare that task with the one of directly measuring the factors for which we are considering the use of indicators. To establish that a species or set of species is an ecological indicator, we need to do considerable experimental work, including comparisons of the species in different localities. This work is to ensure that there is a high correlation between the indicator(s) and the appropriate environmental conditions. Wilber (1969), among others, summarizes some aspects of and literature on indicator organisms.

Thus, although a species or related group of species can be used as indicators of ecological conditions, it is more efficient and reliable to use more general characteristics as indicators. Some examples of this use have been whole sets of species (Ellenberg, 1950), amounts of biologically important substances (such as chlorophyll; see Aruga and Monsi, 1963), biochemical oxygen demand measurements (American Public Health Association, 1975), and bioassay with microorganisms (American Public Health Association, 1975).

E. Species Diversity

The usefulness of species diversity as an indicator of ecosystem or community conditions has been overestimated. First, species diversity is affected by many variables other than pollution in an ecosystem. As previously stressed, a range of measures is necessary to characterize the impact of a factor. No single measure, such as species diversity, should be used alone. Second, it is not the case that high species diversity indicates a healthy environment and vice versa; some natural communities (e.g., salt marshes) have rather low species diversity in "healthy" undisturbed states. Third, we cannot compare species diversities across widely differing taxa nor at different times of the year. Diversities must be calculated for coherent systematic groups, or at least organisms of generally similar trophic and size characteristics (Poole, 1974), and for comparable seasonal stages (e.g., Wilhm and Dorris, 1966).

However, species diversity can be used as one of a spectrum of measures to detect impact of a factor on an ecosystem or community. Several indices of species diversity have been proposed and used (Poole, 1974). The most commonly accepted indices, such as the Shannon–Weaver index H, depend on both the number of species present and the relative abundances of each species. Pielou (1966a,b) discusses in detail the appropriate use of various indices for different types of collections. Poole (1974) and Wilhm and Dorris (1966) review the development and meaning of species diversity indices.

Although a low species diversity index does not necessarily indicate an unhealthy environment, some studies show that pollution and other stresses tend to lower species diversity index values by decreases of the rarer species and increases in numbers of a few species. Barrett (1969) found that an acute insecticide (Sevin) stress temporarily reduced species–numbers diversity in all orders of insects in a grassland (see Fig. 2.2). Wilhm (1967), Wilhm and Dorris (1966), Nash (1975), and Tomkins and Grant (1977) also discuss examples of the finding that diversity indices are lowered from their normal levels by pollution. However, reduced diversity does not always result from disturbances (Odum, 1975; Larsen, 1974), and it should not be used as an index to pollution or disturbance.

Nevertheless, it can be useful to measure species diversity in a field survey. Later comparisons with postimpact species diversity indices at the same sites and at comparable times of the year can be one of many indications of community changes due to the manipulated factor (i.e., not occurring in the control condition). Such comparisons must be strictly controlled (side-by-side comparisons are a great help), since an enormous number of natural variables may affect species diversity indices. This is readily understood when we consider the number of variables that may account for changes in the number of individuals of a single species, let alone a collection of species such as represented in species diversity indices.

It should be noted that, among others, Peet (1975) severely criticizes most relative species diversity indices (e.g., Shannon–Weaver H) as being nearly meaningless because of their high sensitivity to small variations in species numbers caused by sample size or stochastic variation. In light of this criticism, it is perhaps advisable to use simply the number of species present as a simple, objective species richness index, remembering of course that it does not contain any information about the distribution of individuals among species. A wise approach may be to examine both the number of species and the

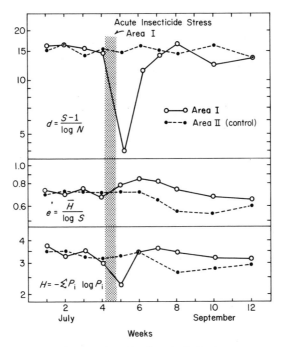

Fig. 2.2. An example of lowered species diversity index in response to an insecticide stress. The insecticide Sevin was applied to a millet field, which was sampled for arthropods (a matched control area was also sampled). The lower graph shows the effects of the stress on a general index of total diversity (*H*, the Shannon–Weaver index). Also shown are the effects on two components of diversity (*d* and *e*). Note, as emphasized in text, that a stress does not always bring about lowered species diversity. (From Odum, 1971.)

calculated indices (see Abele, 1974, for an example; see Peet, 1974, for a guide to indices), if there is a particular interest in the species diversity of the system.

III. SYSTEM FUNCTIONAL STUDY

A. Productivity

One of the most important functional characteristics of ecosystems is their rate of production of organic matter at various trophic levels. Productivity depends to a great extent on the qualitative nature of the ecosystem's components. That is, the set of organisms and environmental conditions existing in the ecosystem determine a characteristic range of productivity values that the system can possess. Thus, in a sense the productivity of an ecosystem, community, or population is a summary of the existing conditions and is one quantitative expression of the physical system's capacity to support living organisms. If an anthropogenic factor alters the productivity of a natural system, it affects the ability of a number of its organisms to maintain their population or biomass levels and thus constitutes an important change that should be measured in an environmental impact study. Of course, the yield of organic matter to humans (e.g., from a fishery) depends heavily on the productivity of the organisms being exploited.

The analysis of productivity in an environmental impact study can take a number of forms, from estimating the overall productivity of the ecosystem to focusing on the productivity of only certain components of interest. In the initial field survey or prestudy, it is important to measure productivity of potential interest for later postimpact comparisons and for selection of those productivities to be studied more fully in a side-by-side comparison scheme. The primary productivity of an ecosystem, community, or species is the rate at which producer organisms, such as green plants, store (produce) organic matter by photosynthesis and/or chemosynthesis. The productivity (rate of production) of a system has the following components:

Gross primary productivity (i.e., total photosynthesis, total assimilation) is the total rate of photosynthesis including that used by respiration during the period of measurement.

Net primary productivity (i.e., apparent photosynthesis or net assimilation) is the gross primary productivity minus the utilization of organic matter by the plants in respiration during the measurement period.

Secondary productivity is the rate of storage (assimilation) of organic matter by a consumer level, minus that used in respiration. Assimilation refers to the total amount of energy flow through the consumer level.

Net community productivity is the net primary production minus the heterotrophic consumption during the measurement period, i.e., the rate of storage of organic matter by the producers of a community in excess of losses to its own autotrophic respiration and to heterotrophic use.

As is probably apparent, since standing crop or biomass (in grams, calories, etc.) is not equivalent to productivity, the rate of production of biomass cannot generally be estimated by measuring the standing crop at any one time. However, standing crop methods are used to estimate productivity in situations where the producing organisms are large and relatively long-lived and immediate utilization of the materials produced is minimal. Where this is not the case, the loss of material produced for the time period in question must also be measured and added to the standing crop or standing biomass.

1. *Primary Productivity*

The methods available to measure productivity in plants (primary productivity) are quite varied. One common method for grasses and forbs is harvesting, in which the increase in biomass from the beginning to the end of a selected time period (often a growing season) is used to estimate the net primary productivity. Sampling is usually done by cropping the above-ground and below-ground plant parts contained in systematically or randomly selected quadrats. The plant parts are weighed, dried, ashed, and reweighed. The productivity is expressed in grams of dry weight per square meter per year, or in kilocalories per square meter per year. Bomb calorimetry or tabled values (Southwood, 1966) are used to measure or calculate caloric content of samples. The production is estimated as the change in biomass plus any losses to herbivores plus any losses through decay of the older plant parts. One difficulty involved, especially when applying the harvest method to perennials, is that one must separate the current year's growth from previous years' materials. Bernard (1974) studied the productivity of a sedge wetland and an adjacent dry old-field using harvest methods. He emphasizes several important points, including that storage and translocation of materials in the plants can lead to errors in estimating a single year's productivity unless winter sampling is performed in addition to that made at the beginning and end of the growing season. Further, most end-of-the-growing-season samples are probably taken too early, since, at least in the plants studied by Bernard, some active growth was still found in November. Further details and examples of the harvest method and its sampling problems are contained in Van Dyne *et al.* (1963), Wiegert (1962), Wiegert and Evans (1964), Lomnicki *et al.* (1968), Andrews *et al.* (1974), and Kirby and Gosselink (1976).

For woody plants, especially larger ones, the harvest method becomes impractical. Methods based on measuring the dimensions of trees and shrubs after an initial calibration using harvest methods are available. Whittaker and Woodwell (1968) discuss these techniques, and Whittaker et al. (1974) apply these methods on a large scale.

Gas exchange methods are also commonly used, primarily for aquatic systems but not uncommonly for terrestrial plant systems as well. In these methods the oxygen produced or the carbon dioxide taken up are measured, since both of these have a definite relationship to the organic matter produced. The oxygen produced or the carbon dioxide taken up are usually measured in enclosed chambers (bottles, spheres, or tents). Strickland and Parsons (1968) and the American Public Health Association's (1975) "Standard Methods" are basic references for these techniques (see also Newbould, 1967; Antia et al., 1963; Woodwell and Whittaker, 1968; Vollenweider, 1965). Some gas exchange methods are possible in unenclosed or partially enclosed systems (Lemon, 1960, 1967; Odum and Pigeon, 1970).

Other commonly used methods to measure primary productivity are uptake of radioactive carbon (Strickland and Parsons, 1968; Thomas, 1964) and change in pH, which depends on the change in dissolved carbon dioxide content (Beyers et al., 1963). Less commonly, methods measuring disappearance of raw materials, chlorophyll content, or remote sensing by aerial photography are used. For particular problems of measuring productivity in plankton, see National Academy of Sciences (1969), Vollenweider (1969), and Lewis (1974).

2. Secondary Productivity

The measurement of secondary productivity is generally difficult and produces crude estimates which are useful primarily to compare magnitudes with other productivities and energy transfers rather than to have any precise idea of the values in question. In this sense, secondary productivity estimates can be useful in environmental impact studies, where we wish to compare productivity relative to a comparable previous estimate or to a comparable estimate in a matched manipulated system. Small differences in such comparisons can be ascribed to error in the estimation procedure, but order of magnitude differences should lead to more detailed studies of the components showing the differences.

In order to estimate the production of an animal population, it is necessary to measure its consumption, respiration, and excretion (energy budget), since production is the net balance between consumption and metabolic use plus excretion. Alternatively and more directly, it is possible to measure the reproduction and growth of the animals in question, since production is due to both of these processes. In either the energy budget or the reproduction and growth approach, the measurements are usually made on a few animals in the laboratory or in the field, and the results are extrapolated to the estimated size of the population (e.g., Van Hook and Dodson, 1974). In making the extrapolations, it is important to keep in mind the fact that a population usually consists of several age classes of varying energy budgets, growth rates, and reproductive rates. Accordingly, separate data for each age class are usually required (for an example, see Randolph *et al.*, 1975) unless only one age class enters into the questions being investigated. In addition, seasonal or other time-dependent variation in these parameters should be kept in mind. Further, energy budget estimates based on laboratory measurements usually need a correction for the animals' activity in a natural setting (e.g., Van Hook and Dodson, 1974). This latter correction is at best a rough approximation. Details of the methods used to estimate production by measuring the growth and reproduction of a population of animals are discussed by Petrusewicz and MacFadyen (1970). Winberg (1971) reviews in detail the methods for estimating production of aquatic animals.

B. Trophic Structure and Energy Flow

The overall function of an ecosystem, as incredibly complex as it is, may be summarized and studied by determining the connections among major components and by estimating the energy processed by each major component. The major components are often trophic groups (groups of organisms performing similar feeding functions), but in some cases individual populations may be studied by this approach. The advantages of energy flow studies are many. First, in spite of the quantitative errors inherent in the extrapolations from the measured data to the estimated energy flows and of the difficulty in gathering enough data to make such estimates reliable, the energy flow approach allows us to examine the whole system as well as some of its parts. The importance of such a broad viewpoint has already been emphasized in Chapter 1. Particularly important is the fact that major functional characteristics or operating laws may not be apparent from more limited detailed studies. The importance to impact analysis of uncovering such major functional relationships is self-evident. Second, using energy flow as a currency allows comparison of broadly different systems, since the numbers or weights of organisms can be converted to energy units and thus be comparable in spite of great differences in sizes and kinds of organisms. Third, the biological importance of various components to overall system function can be determined by estimating the energy flow through those components. Again, such importance measures can be used for comparison even across differing habitat types, sizes of individuals, etc. As previously mentioned, although some species can be very important (e.g., structural dominants) without controlling a large portion of the community energy flow, generally the species with control of major portions of the energy flow largely determine the function and the qualitative character of a community.

The study of energy flow begins by identification of the major food chains or the food web. Some of this work is performed by studying feeding relationships as outlined earlier under the system structural study. From studies of such relationships trophic groups are formed and food web diagrams drawn. Then energy flow through each level or component is calculated. The energy flow (total assimilation) through a trophic level is the production plus the respiration and excretion performed at that level. In calculating energy flow we use data on density, biomass, production, and respiration of the organisms in question. The units used are usually kilocalories per square meter per day or year.

The best known examples of studies of energy flow through ecosystems are probably Odum's (1957) work on Silver Springs, Florida, and Teal's (1962) study of a salt marsh ecosystem in Georgia. The salt marsh study is of particular interest because it represents a relatively complex and ubiquitous ecosystem type. Teal studied the marsh flora and fauna and constructed a food web diagram (see Fig. 2.3). He then made a large number of estimates based on his direct measurements, on other published data, and on inference from other information to calculate the energy flow through each trophic group. He was then able to construct the quantitative community energy flow diagram shown in Fig. 2.4 and the summary table of salt marsh energetics shown in Table 2.1. From this information he was able to conclude, among other things, that the salt marsh is primarily a detritus food chain system and only secondarily a grazing food chain system, that

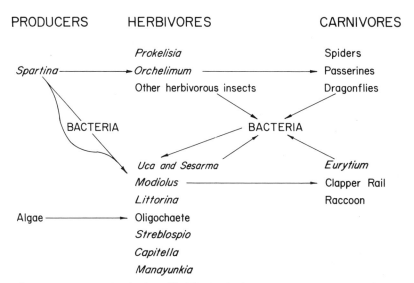

PRODUCERS   HERBIVORES   CARNIVORES

Prokelisia   Spiders

Spartina ————————→ *Orchelimum* ——————————→ Passerines

Other herbivorous insects   Dragonflies

BACTERIA   BACTERIA

Uca and Sesarma   *Eurytium*

Modiolus ——————————————————→ Clapper Rail

Littorina   Raccoon

Algae ———————————→ Oligochaete

Streblospio

Capitella

Manayunkia

Fig. 2.3. An example of a simplified food web diagram. The grouping of organisms with similar food sources constitutes trophic grouping; detailed species studies are often unnecessary if trophic groups are fairly homogeneous. (From Teal, 1962.)

the salt marsh obtains its apparent stability by having species with rather unrestricted diet requirements (i.e., little food specificity), and that the salt marsh exports about 45% of its net production to the estuary by tidal flushing (see Fig. 2.4). Although some of these conclusions have been somewhat modified since Teal's study, it is obvious that Teal was formulating ecologically important hypotheses, and that these were possible because of the breadth of the study. Such hypotheses or conclusions are well worth the sacrifice of precision on many of the detailed studies involved and can be of great importance in an environmental impact study, since we are looking for qualitative characteristics that may be especially sensitive or important and that will, therefore, help us to predict changes in the ecosystem. For example, having seen that detritus feeders are important to the natural function of a salt marsh, we would be well advised to concentrate, in future studies on salt marsh pollution, on detritus feeders in particular (see Section V).

Other examples of ecosystem and population energy flow studies are Kuenzler (1961), Odum and Smalley (1959), Odum and Odum

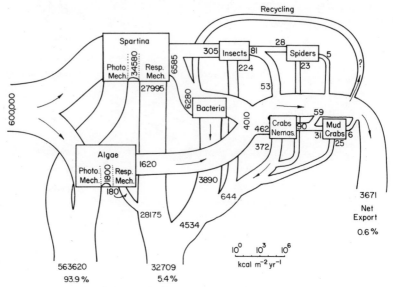

Fig. 2.4. An example of a community energy flow diagram. Note that of the 6585 kcal m^{-2} yr^{-1} net productivity of the grasses (*Spartina* sp.) 6280 kcal m^{-2} yr^{-1} are estimated to go into detrital (decomposer) pathways, while only 305 kcal m^{-2} yr^{-1} are estimated to be eaten alive by herbivorous species. Also note the relatively large estimated net export (3671 kcal m^{-2} yr^{-1}, or 45% of net production) to the estuary. (From Teal, 1962.)

(1955), Smalley (1960), Teal (1957), Tilly (1968), Wiegert (1965), Golley (1960), and Andrews *et al.* (1974).

It should be emphasized that the errors inherent in energy flow estimates must be kept in mind when using the estimates and the conclusions drawn from them. In basing decisions on energy flow studies, questions of the form "what is the kilocalorie per square meter per year assimilation of fiddler crabs?" are not advisable, but questions of the form "since we conclude that the decomposers in this system are qualitatively important, should we perform an experiment on the effects of factor X on the decomposers?" or "how important are the herbivores in system A compared to the herbivores in system B as judged by energy flow control?" can be of great value. It should be noted that energy flow studies can provide a great deal of basic information on functional relationships to be included, along with other information, in mathematical modeling of natural systems.

C. Nutrient Relationships

The functioning and productivity of an ecosystem depend on the kinds and amounts of nutrients available as well as on energy processing. This fact is evident from the marked increases in productivity measured when a limiting nutrient is added to an ecosystem. The nutrient concentrations, pathways, and flux rates in ecosystems are operating characteristics that may change in response to human activities in those systems. Therefore, nutrient studies are often of interest in impact analysis. This is particularly true when nutrient enrichment (e.g., pollution by sewage and/or detergents) of an ecosystem is involved.

Nutrient cycling is similar to energy flow, with nutrients flowing among ecosystem compartments or components. In a rough sense, energy flow is an open system (with input from radiant energy and output in the form of heat dissipation), while nutrient cycling is essentially a closed system (except where inputs and outputs to adjacent ecosystems are possible). Nutrients are described as being in pools, which refer to the quantity of a chemical substance in some component of an ecosystem. An example of a nitrogen pool is the quantity of nitrogen in the zooplankton of a lake. The nutrients are transferred from pool to pool at different rates called flux rates. Flux rates are usually studied in addition to sizes of pools, since turnover rate rather

TABLE 2.1.

Summary Table of Salt Marsh Energetics[a]

Input as light	600,000 kcal/m²/yr
Loss in photosynthesis	563,620 or 93.9%
Gross production	36,380 or 6.1% of light
Producer respiration	28,175 or 77% of gross production
Net production	8205 kcal/m²/yr
Bacterial respiration	3890 or 47% of net production
Primary consumer respiration	596 or 7% of net production
Secondary consumer respiration	48 or 0.6% of net production
Total energy dissipation by consumers	4534 or 55% of net production
Export	3671 or 45% of net production

[a] From Teal (1962).

than concentration determines productivity (analogously to productivity versus standing crop). Turnover rate (the fraction of the total amount of a chemical in a component which leaves or enters in a given length of time) and turnover time (the time needed to replace a quantity of the chemical equal to the amount in the component) are also often estimated. In addition, nutrient studies sometimes estimate residence time (the length of time that a given amount of substance remains in the component being discussed). Such estimates are often made using radiotracers, as discussed by Robertson (1957). For example, Van Hook (1971) used radioisotopes to study the cycling of potassium, calcium, and sodium in a grassland ecosystem. Other methods commonly used for nutrient analysis are spectrophotometric methods, chemical methods such as Kjeldahl techniques for nitrogen, and use of instruments called elemental analyzers. Strickland and Parsons (1972) is a standard reference on methods for nutrient analysis; Glass (1973) contains several papers on nutrient methods as well.

When a nutrient shows a high assimilation rate (i.e., a relatively high proportion of the available nutrient is used) compared to other nutrients, that nutrient is probably a limiting factor. Determination of limiting nutrients can be of significance to some applied problems. For example, Ryther and Dunstan (1971) showed that nitrogen, rather than phosphorus, is the limiting factor to algal growth and eutrophication in coastal marine environments. Thus, they concluded that removal of phosphates from detergents to alleviate coastal eutrophication would not be effective. Further, replacement of phosphates with nitrogen-containing NTA in detergents would probably worsen the eutrophication effect.

Alterations in nutrient relationships may not only affect the productivity of ecosystems, but may also have dramatic impacts on species composition. For example, the establishment of duck farms in the rivers near Great South Bay, Long Island, greatly enriched the waters in the bay by addition of duck manure. A considerable increase in phytoplankton density resulted. Such an increase in productivity appeared at first to be desirable. However, the phytoplankton composition changed (Ryther, 1954) from the normal mixture of diatoms, green flagellates, and dinoflagellates to almost all small green flagellates (*Nannochloris* and *Stichococcus*, neither of which was dominant previously). The shift in species composition was apparently because the added nutrients were in organic form (which the new species could readily utilize but which the "normal" species could not) and because there was a low nitrogen to phosphorus ratio (which also favored the new dominant species). The change in phytoplankton species composition had further dramatic impacts. The area had previously supported a considerable oyster population, which was used commercially. When the phytoplankton composition changed, the oysters and other shellfish starved to death in spite of consuming the small green flagellates in large quantities. The oysters were apparently unable to digest the new phytoplankton species. The oysters and several other shellfish disappeared from the area, and reintroduction attempts failed. This example illustrates the far-reaching impacts that changes in nutrient relationships can have and also the subtlety and complexity of ecosystem interactions. Investigation of the fertilization effects using an ecological approach, previous to the establishment of the duck farms, could have allowed prediction of the effect on the shellfish populations. Note, of course, that the duck farms may still have been considered to be more "valuable" in some sense than the oyster industry, and the outcome may have been the same. However, the decision could have been made in knowledge of the probable outcome.

It should be noted that microorganisms, such as bacteria, play important roles in biogeochemical cycles and therefore in nutrient cycles. In particular, the oxygen, carbon, nitrogen, and sulfur cycles are heavily dependent on bacterial action (Odum, 1971; Brock, 1966, 1974; Veldkamp, 1975).

D. Decomposition Processes

In the study of energy flow of an ecosystem, the major processes investigated are primary production, energy processing by consumers, and decomposition. Perhaps because of the methodological difficulties involved in the study of decomposition and the microorganisms that play a large part in this process, ecological impact studies often omit study of the breakdown of organic matter and its responses to changes in the environment. However, decomposition processes are vital to ecosystem function and should be considered in impact analyses.

Decomposition of organic matter results from physical activities, such as fires or tidal action, and biological activities, such as bacterial and fungal action. Bacteria and fungi are the principal decomposers in most ecosystems, but other organisms, such as protozoa, nematodes, and small arthropods, are also important decomposers (Tribe, 1957, 1961; Johannes, 1965; Crossley and Witkamp, 1964). In many systems, detritivores (organisms feeding largely on decomposing particulate plant and animal material containing microorganisms) are important to decomposition processes (Melchiorri-Santolini and Hopton, 1972; Teal, 1962). Detritivores accelerate breakdown of resistant plant materials by breaking it into smaller fragments, thereby creating more surface area available for microbial attack. Further, detritivores may stimulate bacterial growth by adding stimulatory proteins or growth substances and by keeping populations at levels conducive to rapid growth rate (Odum, 1971). Thus, decomposition is a complex process involving a variety of organisms and varying considerably from ecosystem to ecosystem. A number of valuable papers on decomposition in a variety of ecological systems are presented in a recent volume edited by Anderson and MacFadyen (1976).

Two general approaches are evident to assess the effects of a factor on decomposition processes in an ecosystem. We may determine which organisms are principal decomposers in the system and test the effects of the factor on those organisms *in situ* and perhaps in laboratory studies. This approach is useful when larger organisms, such as some detritivores, are involved, but perhaps not practical for analyzing effects on mixed bacterial and fungal communities. It is possible to determine the types, numbers, and biomass of microorganisms in an ecosystem (Wiebe, 1971), but these determinations present considerable problems and the resulting information is of dubious value. It is more important to measure the total activity of a microbial population. This can be done by determining carbon dioxide production or oxygen consumption or by measuring rates of substrate decomposition. Details and discussion of these methods can be found in Brock (1966), Wiebe (1971), Hobbie (1971), Hobbie *et al.* (1968), Coleman (1973), and Klein (1972). Andrews *et al.* (1974) estimated microbial activity by respirometry and the assumed turnover time of the standing crop. A particularly applicable method for impact analysis is the "litter bag" method for study of rate of decomposition of the decomposing material (Odum and de la Cruz, 1967; Crossley and Hoglund, 1962). "Litter bags" are made of mesh, usually nylon or fiberglass, and contain some of the litter or other material that undergoes decomposition in the natural system. The bags are placed in the field and weighed periodically to obtain a measure of decomposition rate. Although litter bags often produce absolute estimates with considerable errors, they are useful for comparative studies. If the litter is labeled with a radioactive tracer, mineral release can also be measured (Odum, 1971, p. 373). And, of course, organisms can be removed from the bags to determine decomposer organisms present. Thus litter bag methods can be used in impact analysis to test for the effects of a factor on decomposition rate, mineral release rate, and kinds and numbers of decomposer organisms present when compared to a control condition.

Several facts should be noted with respect to study of decomposition processes. First, temperature and moisture are important regulators of decomposition processes. This should be kept in mind when testing impact of a factor on decomposition rates. There may well be interactions between the impacting factor and temperature and moisture. Second, when the impacting factor is the addition of a substance, it is important to consider whether the substance is slow to decompose in the environment (Payne et al., 1970) and/or whether the substance is likely to specifically affect important decomposers. (For example, an insecticide is likely to be detrimental to soil microarthropods such as mites and collembolans.) Third, it should be noted that microorganisms, such as bacteria and fungi, should be considered in impact analysis not only as decomposer organisms but also for their important roles in primary production and in biogeochemical cycles (Brock, 1966; Odum, 1971; Veldkamp, 1975; Alexander, 1975). It is of note that Harte and Levy's (1975) analysis predicts that "damage to the decomposers or the organic or inorganic nutrient pools in an ecosystem is a potential source of instability—greater, perhaps, than that arising from tampering with the more visible predator–prey components of the system."

E. Succession or Development of Communities

Succession is a very complex and much debated phenomenon. Succession refers to a reasonably directional change in species structure of a community with time. General discussions of succession can be found in Poole (1974) and Krebs (1972). Succession of communities has been described systematically and in a general way by two distinct schemes. Margalef (1963, 1968) describes community development by the concept of maturity, summarized in Table 2.2, while Odum (1969)

TABLE 2.2.

Summary of Margalef's Model of Ecological Succession ("Maturity")[a]

Characteristic	Ecosystem condition	
	Less mature	More mature
Structure		
Biomass	Small	Large
Species diversity	Low	High
Stratification	Less	More
Energy flow		
Food chains	Short	Long
Primary production per unit of biomass	High	Low
Individual populations		
Fluctuations	More pronounced	Less pronounced
Life cycles	Simple	Complex
Feeding relations	Generalized	Specialized
Size of individuals	Smaller	Larger
Life span of individuals	Short	Long
Population control mechanisms	Abiotic	Biotic
Exploitation by man		
Potential yield	High	Low
Ability to withstand exploitation	Good	Poor

[a] As ecosystems mature, ecosystem features change as indicated. From Krebs (1972).

uses the concept of succession, summarized in Table 2.3. Whichever concept is used, community development can be affected by pollution and other anthropogenic factors. In many cases, succession is reversed under stress conditions; this is termed retrograde succession or retrogression. Such retrogression has been described in an oak–pine forest subjected to chronic γ irradiation (Woodwell, 1970), in tropical forests in Vietnam subjected to chronic herbiciding (Tschirley, 1969), and in other systems.

In an environmental impact field survey, information from the literature or survey data taken on the habitat of interest may indicate a successional trend in the system. If a clear indication of succession exists, we can plan to perform experiments to determine if the proposed impacting factor will alter or reverse the trend. This would be especially pertinent if the communities involved are undergoing relatively rapid succession.

TABLE 2.3.

Summary of Odum's Model of Ecological Succession[a]

Ecosystem attributes	Developmental stages	Mature stages
Community energetics		
1. Gross production/community respiration (P/R ratio)	Greater or less than 1	Approaches 1
2. Gross production/standing crop biomass (P/B ratio)	High	Low
3. Biomass supported/unit energy flow (B/E ratio)	Low	High
4. Net community production (yield)	High	Low
5. Food chains	Linear, predominantly grazing	Weblike, predominantly detritus
Community structure		
6. Total organic matter	Small	Large
7. Inorganic nutrients	Extrabiotic	Intrabiotic
8. Species diversity—variety component	Low	High
9. Species diversity—equitability component	Low	High
10. Biochemical diversity	Low	High
11. Stratification and spatial heterogeneity (pattern diversity)	Poorly organized	Well-organized
Life history		
12. Niche specialization	Broad	Narrow
13. Size of organism	Small	Large
14. Life cycles	Short, simple	Long, complex
Nutrient cycling		
15. Mineral cycles	Open	Closed
16. Nutrient exchange rate, between organisms and environment	Rapid	Slow
17. Role of detritus in nutrient regeneration	Unimportant	Important
Selection pressure		
18. Growth form	For rapid growth ("r selection")	For feedback control ("K selection")
19. Production	Quantity	Quality
Overall homeostasis		
20. Internal symbiosis	Undeveloped	Developed
21. Nutrient conservation	Poor	Good
22. Stability (resistance to external perturbations)	Poor	Good
23. Entropy	High	Low
24. Information	Low	High

[a] Note general similarity to Margalef's model. From Odum (1969). Copyright 1969 by the American Association for the Advancement of Science.

F. Individual Species Characteristics

In discussing study of individual species in environmental impact analysis, it should be noted that many impact studies will be species oriented rather than ecosystem or community oriented. Single species-oriented studies can be performed with an ecological systems orientation, as illustrated in Chapter 6. Even within the framework of ecosystem and community analysis, studies on individual species may often be necessary to investigate overall system responses. Often, characteristics of individual species determine the overall response. A good example of this phenomenon is the case of the duck farms, phytoplankton, and oysters described in Section III,C. The particular nutrient utilization capabilities of the phytoplankton species completely changed the qualitative (and, of course, the quantitative) nature of the bay community.

Often, an anthropogenic change will cause unexpected impacts because of particular characteristics of a species. For example, some of the fish kills caused by shutdown of coastal nuclear power plants in the winter in the northeastern United States were not understood until the seasonal migration behavior of the fish in question was known. Such fish species normally move away from these coastal areas and into deeper waters or warmer latitudes as winter approaches. However, the presence of the heated effluent inhibits movement away from the area. The fish then spend the winter in the warmer water areas around the nuclear plant effluents. For example, bluefish normally move offshore by the time the water reaches about 6°C. However, bluefish will not move offshore if in the vicinity of a heated plume (Raney, 1972, p. 125). When malfunction or routine maintenance of the plant caused shut off of heated effluent, the fish were rather suddenly exposed to the considerably colder water flushing in with the tides and died, sometimes in large numbers. Such kills, of course, affect the entire community because the fish are not only removed as predators and so on but also and primarily because of their fouling effect.

As indicated above, a behavioral characteristic of a component species may strongly influence or determine the response of a community to an impacting factor. Similarly, other functional characteristics (predator–prey relationships, competition, dispersal, adaptation, and others too numerous and complex to be discussed individually here) of a particular species in a community may be of great importance to overall community response. Thus, in any field survey, it is of obvious value to observe such functional traits and relationships qualitatively, at least of those species that appear important in some sense (dominant, key species, etc.) and to incorporate such relationships into verbal models of community function. In short, we want to use species characteristics we observe to increase our understanding of ecological interactions of the species with the rest of its community or ecosystem.

IV. TEMPORAL CHANGES IN STRUCTURAL AND FUNCTIONAL CHARACTERISTICS

It must be emphasized that the field/literature survey constitutes only a beginning, a tentative base of information upon which to formulate hypotheses. It cannot disclose the changes that will obviously occur naturally with time in the undisturbed system. Therefore, it is important to (1) continue the data gathering, in whatever form is decided (see Section V), after the field survey period; (2) couple the survey study with information from later experiments, particularly with the control condition of the side-by-side comparison scheme (see Chapter 1, Section VI,B); and (3) gather data on the structural and functional parameters over the seasons.

V. DECISION MAKING AFTER THE SURVEY

The field survey, if done judiciously, should yield a good general picture of the structure and function of the system being studied. This information should then be used in model building, as discussed in Chapter 3. The modeling process is a way of deciding which aspects of the system need further investigation in order to construct the model. However, the results of the field survey alone can also be considered as a means of deciding what aspects of the system should be investigated experimentally. We can then more intelligently select aspects of the system that may show marked changes when affected by the proposed factor. First, we must decide which major ecological characteristics to test against the impacting factor. For example, considering species diversity as a possible test characteristic, our Rutgers salt marsh research group (see Chapter 6, Section III) decided not to examine the effects of insecticides on the species diversity of crustaceans. Since this diversity was quite low naturally, we would not expect to be able to detect any marked changes due to insecticide treatments. However, the species diversity of marsh insects was considerable, and it seemed probable that this parameter would be affected by the treatments; therefore, this was one of the many variables selected to test against the treatments.

In addition to considering the various system characteristics as candidates for testing against the proposed factor, we should decide which species should be examined for possible impact. In this selection, we can concentrate on the most "important" species, where importance can be defined in all of the following ways:

1. Importance in terms of the species' role in the community. Is this a dominant species? For example, our research group agreed that we should look at whether insecticides affected the productivity of *Spartina* sp. on the salt marsh; there was no reason to believe that it would, physiologically or otherwise, but the role of *Spartina* on the marsh as the major primary producer meant that any effect on this species may have considerable consequences for the whole community.

2. Importance in terms of suspected susceptibility to the proposed change. There are usually obvious or subtle indications of susceptibility, based on the mode of action or structure of the proposed change, which can guide us in picking the species to examine for possible impact. For example, in assessing the possible effects of insect juvenile hormone analogs (third generation pesticides) on an ecosystem, it would seem logical to test any important nontarget insects and crustaceans in the community. This is because the hormone analogs interact with the molting hormonal systems of the target insect and are thus likely to interact with similar systems present in related organisms (other insects, crustaceans). It is much more likely that the hormone analogs will affect nontarget insects and crustaceans, rather than organisms that do not possess similar hormones (e.g., plants, vertebrates).

3. Importance in human or immediate economic terms. We will probably want to test whether a proposed change will have an impact, for example, on a commercially valuable fish as opposed to a noncommercially used fish given that other factors, such as biological importance, are equal. Importance may also be ascribed to possible

human impacts because of aesthetic, as well as economic, considerations.

4. Importance in nondirect relationship to the proposed change. This category covers a number of phenomena. One example is to test whether the proposed change (e.g., insecticide use) will adversely affect some natural mechanism (e.g., killifish's ability to prey on mosquito larvae) which independently achieves a desired end (e.g., mosquito control). Thus, on a salt marsh, even if we did not consider killifish important from the points of view of community function, suspected susceptibility, or human terms, we will be interested in examining these fish because they constitute a potentially valuable independent mosquito control mechanism which may be adversely affected by the proposed insecticide treatment.

Thus, we must consider the following as possibilities for study of biological impact by the proposed change:

a. Aspects shown by modeling to need investigation (see Chapter 3)
b. Functional and structural ecosystem or community characteristics
c. "Important" species (at least four kinds of possible importance)

Note that there may be overlap in these categories. For example, in our salt marsh study the grass *Spartina alterniflora* was selected for study because (1) it was responsible for most of the marsh's primary productivity, the latter being a major functional characteristic, and (2) it was an "important" species because it is a marsh community dominant species.

REFERENCES

Abele, L. G. (1974). *Ecology* **55**, 156–161.
Alexander, M. (1975). *In* "Unifying Concepts in Ecology" (W. H. van Dobben and R. H. Lowe-McConnell, eds.), pp. 224–229. Junk, The Hague.
American Public Health Association (1975). "Standard Methods for the Examination of Water and Wastewater Including Bottom Sediments and Sludges," 14th Ed. APHA, New York.
Anderson, J. M., and MacFadyen, A., eds. (1976). "The Role of Terrestrial and Aquatic Organisms in Decomposition Processes." Blackwell, Oxford.

Andrews, R., Coleman, D. C., Ellis, J. E., and Singh, J. S. (1974). *Proc. Int. Congr. Ecol.*, *1st, The Hague* pp. 22–38. Cent. Agric. Publ., Doc., Wageningen.

Antia, N. J., McAllister, C. D., Parsons, T. R., Stephens, K., and Strickland, J. D. H. (1963). *Limnol. Oceanogr.* **8**, 166–183.

Aruga, Y., and Monsi, M. (1963). *Plant Cell Physiol.* **4**, 29–39.

Barrett, G. W. (1969). *Ecology* **49**, 1019–1035.

Bernard, J. M. (1974). *Ecology* **55**, 350–359.

Beyers, R. J., Larimer, J., Odum, H. T., Parker, R. B., and Armstrong, N. E. (1963). *Publ. Inst. Mar. Sci., Univ. Tex.* **9**, 454–489.

Breen, P. A., and Mann, K. H. (1976). *Mar. Biol.* **34**, 137–142.

Brock, T. D. (1966). "Principles of Microbial Ecology." Prentice-Hall, Englewood Cliffs, New Jersey.

Brock, T. D. (1974). "Biology of Microorganisms." Prentice-Hall, Englewood Cliffs, New Jersey.

Coleman, D. C. (1973). *Oikos* **24**, 361–366.

Crossley, D. A., and Hoglund, M. P. (1962). *Ecology* **43**, 571–573.

Dayton, P. K. (1971). *Ecol. Monogr.* **41**, 351–389.

Dayton, P. K. (1975). *Ecol. Monogr.* **45**, 137–159.

Dempster, J. P. (1960). *J. Anim. Ecol.* **29**, 149–167.

Eliassen, R. (1952). *Sci. Am.* **186**, 17–21.

Ellenberg, H. (1950). "Landwistschaftliche Pflanzensoziologie," Band 1, "Unkrautgemeinschaftenals Zeiger für Klima und Boden." Ulmer, Stuttgart.

Glass, G. E., ed. (1973). "Bioassay Techniques and Environmental Chemistry." Ann Arbor Sci. Publ., Ann Arbor, Michigan.

Golley, F. B. (1960). *Ecol. Monogr.* **30**, 187–206.

Harte, J., and Levy, D. (1975). *In* "Unifying Concepts in Ecology" (W. H. van Dobben and R. H. Lowe-McConnell, eds.), pp. 208–223. Junk, The Hague.

Heald, E. J. (1969). The production of organic detritus in a south Florida estuary. Ph.D. Thesis, Univ. of Miami, Coral Gables, Florida.

Hobbie, J. E. (1971). *In* "The Structure and Function of Fresh-Water Microbial Communities" (J. Cairns, Jr., ed.), Res. Div. Monogr. No. 3, pp. 181–194. VPI, Blacksburg, Virginia.

Hobbie, J. E., Crawford, C. C., and Webb, K. L. (1968). *Science* **159**, 1963–1964.

Holling, C. S. (1973). *Annu. Rev. Ecol. Syst.* **4**, 1–23.

Jeffries, H. P. (1972). *Limnol. Oceanogr.* **17**, 433–440.

Johannes, R. E. (1965). *Limnol. Oceanogr.* **10**, 434–442.

Kirby, C. J., and Gosselink, J. G. (1976). *Ecology* **57**, 1052–1059.

Klein, D. A. (1972). "Systems Analysis of Decomposer Functions in the Grassland Ecosystem," US/IBP Grassland Biome Tech. Rep. No. 201. Colorado State Univ., Fort Collins.

Krebs, C. J. (1972). "Ecology. The Experimental Analysis of Distribution and Abundance." Harper, New York.

Kuenzler, E. J. (1961). *Limnol. Oceanogr.* **6**, 191–204.

Larkin, P. A. (1971). *J. Fish. Res. Board Can.* **28**, 1493–1502.

Larsen, P. F. (1974). *Proc. Int. Congr. Ecol., 1st, The Hague* pp. 80–85. Cent. Agric. Publ. Doc., Wageningen.

Lemon, E. R. (1960). *Agron. J.* **52**, 697–703.

Lemon, E. R. (1967). *In* "Harvesting the Sun" (A. San Pietro, F. A. Greer, and T. J. Army, eds.), pp. 263–290. Academic Press, New York.

Lewis, W. M., Jr. (1974). *Ecol. Monogr.* **44**, 377–409.

Lomnicki, A., Bandola, E., and Jankowska, K. (1968). *Ecology* **49**, 147–149.

Margalef, R. (1963). *Adv. Front. Plant Sci. (Inst. Adv. Sci. Cult., New Delhi)* **2**, 137–188.

Margalef, R. (1968). "Perspectives in Ecological Theory." Univ. of Chicago Press, Chicago, Illinois.

Marples, T. G. (1966). *Ecology* **47**, 270–277.

Melchiorri-Santolini, U., and Hopton, J. W., eds. (1972). *Mem. Ist. Ital. Idrobiol.* **29**, Suppl.

Nash, T. H., III (1975). *Ecol. Monogr.* **45**(2), 183–198.

National Academy of Science (1969). *Eutrophication: Causes, Consequences, Correctives, Int. Symp. Eutrophication, Washington, D.C.*

Neal, B. R., Pulkinen, D. A., and Owen, B. D. (1973). *Can. J. Zool.* **51**(7), 715–721.

Neave, F. (1953). *J. Fish. Res. Board. Can.* **9**, 450–491.

Newbould, P. J. (1967). "Methods of Estimating the Primary Production of Forests," IBP Handb. No. 2. Blackwell, Oxford.

Odum, E. P. (1969). *Science* **164**, 262–270.

Odum, E. P. (1971). "Fundamentals of Ecology," 3rd Ed. Saunders, Philadelphia, Pennsylvania.

Odum, E. P. (1975). *In* "Unifying Concepts in Ecology" (W. H. van Dobben and R. H. Lowe-McConnell, eds.), pp. 11–14. Junk, The Hague.

Odum, E. P., and de la Cruz, A. A. (1967). *In* "Estuaries" (G. Lauff, ed.), Publ. No. 83, pp. 383–388. Am. Assoc. Adv. Sci., Washington, D.C.

Odum, E. P., and Kuenzler, E. J. (1963). *In* "Radioecology" (V. Schultz and A. W. Klement, eds.), pp. 113–120. Reinhold, New York.

Odum, E. P., and Smalley, A. E. (1959). *Proc. Natl. Acad. Sci. U.S.A.* **45**, 617–622.

Odum, H. T. (1957). *Ecol. Monogr.* **27**, 55–112.

Odum, H. T., and Odum, E. P. (1955). *Ecol. Monogr.* **25**, 291–320.

Odum, H. T., and Pigeon, R. F., eds. (1970). "A Tropical Rain Forest. A Study of Irradiation and Ecology at El Verde, Puerto Rico." Natl. Tech. Inf. Serv., Springfield, Virginia.

Odum, W. E. (1970). Pathways of energy flow in a south Florida estuary. Ph.D. Thesis, Univ. of Miami, Coral Gables, Florida.

Paine, R. T. (1969). *Am. Nat.* **103**, 91–93.

Paine, R. T. (1974). *Oecologia* **15**, 93–120.

Patrick, R. (1965). *In* "Third Seminar on Biological Problems in Water Pollution," No. 999-WP-25, pp. 225–230. U.S. Public Health Serv., Cincinnati, Ohio.

Payne, W. J., Wiebe, W. J., and Christian, R. R. (1970). *BioScience* **20**, 862–865.

Peet, R. K. (1974). *Annu. Rev. Ecol. Syst.* **5**, 285–308.

Peet, R. K. (1975). *Ecology* **56**, 496–498.

Petrusewicz, K., and MacFadyen, A. (1970). "Productivity of Terrestrial Animals: Principles and Methods," IBP Handb. No. 13. Blackwell, Oxford.

Pielou, E. C. (1966a). *J. Theor. Biol.* **10**, 370–383.

Pielou, E. C. (1966b). *Am. Nat.* **100**, 463–465.

Poole, R. W. (1974). "An Introduction to Quantitative Ecology." McGraw-Hill, New York.

Randolph, P. A., Randolph, J. C., and Barlow, C. A. (1975). *Ecology* **56**, 359–369.
Raney, E. C. (1972). "Ecological Considerations for Ocean Sites off New Jersey for Proposed Nuclear Generating Stations," Vol. I, Part 2. Icthyological Associates, Ithaca, New York.
Rawson, D. S. (1956). *Limnol. Oceanogr.* **1**, 18–25.
Reynoldson, T. B., and Young, J. O. (1963). *J. Anim. Ecol.* **32**, 175–191.
Robertson, J. S. (1957). *Physiol. Rev.* **37**, 133–154.
Ryther, J. H. (1954). *Biol. Bull. (Woods Hole, Mass.)* **106**, 198–209.
Ryther, J. H., and Dunstan, W. M. (1971). *Science* **171**, 1008–1012.
Shure, D. J. (1970). *Ecology* **51**, 899–901.
Smalley, A. E. (1960). *Ecology* **41**, 672–677.
Southwood, T. R. E. (1966). "Ecological Methods, with Particular Reference to the Study of Insect Populations." Methuen, London.
Stockner, J. G., and Benson, W. W. (1967). *Limnol. Oceanogr.* **12**, 513–532.
Strickland, J. D. H., and Parsons, T. R. (1968). *Bull., Fish. Res. Board. Can.* No 167.
Strickland, J. D. H., and Parsons, T. R. (1972). *Bull., Fish. Res. Board Can.* No. 167 (2nd Ed.).
Sutherland, J. P. (1974). *Am. Nat.* **108**, 859–873.
Teal, J. M. (1957). *Ecol. Monogr.* **27**, 283–302.
Teal, J. M. (1962). *Ecology* **43**, 614–624.
Thomas, W. H. (1964). *U.S. Fish Wildl. Serv., Fish. Bull.* **63**, 273–292.
Tilly, L. J. (1968). *Ecol. Monogr.* **38**, 169–197.
Tomkins, D. J., and Grant, W. F. (1977). *Ecology* **58**, 398–406.
Tribe, H. T. (1957). *In* "Microbial Ecology" (R. E. O. Williams and C. C. Spicer, eds.), Symposium of the Society for General Microbiology, Vol. 7, pp. 287–298. Cambridge Univ. Press, London and New York.
Tribe, H. T. (1961). *Soil Sci.* **92**, 61–77.
Tschirley, F. H. (1969). *Science* **163**, 779–786.
Van Dyne, G. M., Vogel, W. G., and Fisser, H. G. (1963). *Ecology* **44**, 746–759.
Van Hook, R. I. (1971). *Ecol. Monogr.* **41**, 1–26.
Van Hook, R. I., and Dodson, G. J. (1974). *Ecology* **55**, 205–207.
Veldkamp, H. (1975). *In* "Unifying Concepts in Ecology" pp. 44–49. (W. H. van Dobben and R. H. Lowe-McConnell, eds.), Junk, The Hague.
Vollenweider, R. A. (1965). *In* "Primary Productivity in Aquatic Environments" (C. R. Goldman, ed.), pp. 427–457. Univ. of California Press, Berkeley.
Vollenweider, R. A., ed. (1969). "A Manual on Methods for Measuring Primary Productivity in Aquatic Environments." Blackwell, Oxford.
Whittaker, R. H. (1965). *Science* **147**, 250–260.
Whittaker, R. H., and Woodwell, G. M. (1968). *J. Ecol.* **56**, 1–25.
Whittaker, R. H., Bormann, F. H., Likens, G. E., and Siccama, T. G. (1974). *Ecol. Monogr.* **44**, 233–252.
Wiebe, W. J. (1971). *In* E. P. Odum, "Fundamentals of Ecology," 3rd Ed., pp. 484–497. Saunders, Philadelphia, Pennsylvania.
Wiegert, R. G. (1962). *Ecology* **43**, 125–129.
Wiegert, R. G. (1965). *Oikos* **16**, 161–176.
Wiegert, R. G., and Evans, F. C. (1964). *Ecology* **45**, 49–63.
Wilber, C. G. (1969). "The Biological Aspects of Water Pollution." Thomas, Springfield, Illinois.

Wilhm, J. L. (1967). *J. Water Pollut. Control Fed.* **39**, 1673–1683.

Wilhm, J. L., and Dorris, T. C. (1966). *Am. Midl. Nat.* **76** 427–449.

Winberg, G. G. (1971). "Methods for the Estimation of Production of Aquatic Animals." Academic Press, New York.

Woodwell, G. M. (1970). *Science* **168**, 429–433.

Woodwell, G. M., and Whittaker, R. H. (1968). *Am. Zool.* **8**, 19–30.

Young, J. O., Morris, I. G., and Reynoldson, T. B. (1964). *Arch. Hydrobiol.* **60**, 366–373.

3

Modeling the System

I. OBJECTIVES

The average biologist performing an environmental impact study does not have experience in mathematical modeling. Many reference works on modeling are available, but almost all presentations are simply too complex for a first exposure; such presentations often discourage attempts to model biological systems. Therefore, here I attempt a greatly simplified step-by-step presentation that should allow any biologist to write down a starting model for his/her system. Once this first step is taken, more involved topics are easier to pursue in the methodological references given and in the examples cited. The main point of this chapter is to overcome the initial hurdle in modeling.

Following an initial overview of a biological system (such as a survey including literature and/or experimental information), the researcher has a mental model of the system. This model or representation of the real system may be only conceptual or may develop into a full-scale computerized mathematical model. Constructing a formal mathematical model has the advantage of forcing a more precise and careful description of the system than conceptual models. As a result of the precise description entailed, formal models can help to guide research or outline a problem for more careful and relevant experimental work than is generally possible using only mental conceptual models. Thus, even if computer facilities and advanced knowledge of mathematics are not available, the step of building a formal explicit model in rough form can be of value in an ecological study. Such a model can

be composed of explicit statements of how system components probably interact and of the order in which these interactions take place.
The interactions can then be represented as graphs or simple
equations.

In addition to the function of forcing precise explicit statements,
full-scale formal models, usually in computerized form, can be used
for prediction of dynamic changes in the system. In this role, models
produce tentative answers to questions about general system processes. However, models often fail to predict the measured system
responses. In such cases, the model is frequently useful in pointing
out errors in the concepts used to develop the model. New or altered
models can then be constructed. In addition, computerized
mathematical models can be used to explore possible responses of the
system under conditions that were not present empirically in the past
but that may be imagined to arise in the future. Computer experimentation of this kind can be used to do experiments leading to
system destruction or disappearance. A computerized mathematical
model can be used to investigate the possible consequences of many
options rapidly because computers can handle vast amounts of data
rapidly and do so at a relatively low cost. Thus mathematical models
have many advantages that should be considered regardless of the
possibility of predictive failure (inaccuracy). In environmental impact
analysis, precision of system description and exploration of options, as
well as predictive power, are very important. Therefore, mathematical models have great applicability in environmental impact analysis.

II. ABOUT MODELS AND MODELING

Mathematical models of many kinds exist, and this variety adds to
the confusion of an initial exposure to modeling. Pielou (1972) presents a helpful classification of models as follows. Models may (1) treat
time as continuous or discrete, (2) be analytic or simulation models,
(3) be deterministic or stochastic, (4) be deductive or inductive. In
what follows I will briefly clarify these distinctions.

Continuous time treatments are written in terms of differential
equations; such models are often difficult to construct and difficult to
solve. In addition, in many biological systems we study variables that
can only take on discrete values, rather than any value on a continuous numerical scale. Nevertheless, differential equations are used in

many biological models. Watt (1968) presents a scheme for choosing a differential equation to describe a relationship between two variables. Alternately, models may use discrete time bases in formulating the rules for change; these rules are then stated as difference equations. The essential distinction between these types of equations is that variables and time change only stepwise in difference equations, but continuously in differential equations. [Watt (1968) elaborates on this distinction.] Difference equations used in biological models have been either event oriented or state oriented. In event-oriented difference equations we start by fixing the amount of change wanted in a variable and then calculate the amount of time needed for this change to take place (e.g., Holling, 1965, 1966). Much more commonly, the difference equations used are state oriented. State variables are those properties that describe the state of a (biological) system, such as numbers of animals, amount of nutrients, biomass, etc. Using state-oriented difference equations, we start with a list of state variable values at a given time and calculate the state variable values at a fixed later time. State-oriented fixed-interval difference equations are usually the easiest to specify and are widely used; these will be used in this presentation.

Models may also be of the analytic or the simulation types. In analytic models, algebraic and other mathematical manipulations are used to investigate the results of the equation systems (models). This is an enormous field of study that usually involves knowledge of higher level mathematics. Usually, analytical models become mathematically intractable if they include biologically realistic complex sets of assumptions. In simulation models, computers are used to investigate the outcomes of the equation systems. Simulation models have the advantages of not requiring knowledge of higher level mathematics and of being well suited to handling large numbers of realistic assumptions. These features make simulation models extremely valuable in environmental impact studies.

Simulation models may be deterministic, leading to only one possible outcome for each set of state variable values, or stochastic, incorporating random processes and thus estimating the expected variability of the model results (outputs). The methods used to incorporate random processes are called Monte Carlo methods and are discussed concisely by Poole (1974) and Watt (1968). It should be emphasized that stochastic models may produce different results from repeated runs with the *same* starting values for the state variables.

Finally, models may be deductive or inductive. Deductive models are constructed by stating logical hypotheses about processes in the system and then comparing the model to data. Inductive or empirical models are formulated by examining experimental data and establishing algebraic relationships from the data. The latter type requires rather precise and accurate data, does not usually provide much understanding of the processes involved, and may easily lose predictive power outside the normal (measured) range of conditions. Nevertheless, both types of models and hybrids of these types are common, and both can be useful for different purposes.

III. STEP-BY-STEP MODELING

In this section, an example of a biological impact analysis problem will be used to illustrate a modeling effort. The model will be very simple, but in many ways typical of the problems encountered in modeling impacted biological systems. The example is the study of the effects of the organophosphorus insecticide temephos (Abate) on fiddler crab populations inhabiting salt marshes sprayed with temephos for mosquito control. Note that a model could focus attention at a number of possible levels, including the overall marsh with all its component species. However, for a number of reasons, one of the levels chosen for modeling was the population of fiddler crabs. Note that this model could also be considered as a submodel of a larger community model and be used as such. The question we are seeking to answer is whether and to what extent fiddler crab populations are affected by temephos treatment of the marsh. By using modeling to approach the problem, we hope to (1) be able to predict fiddler crab population changes that may occur, (2) uncover unanswered questions and possible conceptual errors that are important to describe the system precisely, (3) investigate and predict consequences of using and not using the insecticide, as well as of using it at different rates and concentrations. Note that it is important to do modeling early in the research project, since modeling is very useful in pointing out questions that need investigation as well as indicating aspects that can be neglected as unimportant to model building and testing. Thus, empirical research can be much better directed and less wasteful when coupled with modeling from the beginning.

In the step-by-step presentation that follows, the general step will be described first and the application of that step to the fiddler crab problem will follow. It should be noted that these are my adaptations of some specific approaches used by Holling (Munn, 1975, Chapter 5; Clark *et al.*, 1978) and Walters (Walters and Efford, 1972; Walters *et al.*, 1975), as well as of more generalized approaches, such as those described in Patten (1971), Watt (1966), and Peppard (1975).

A. Problem Specification and Bounding

The first step in constructing a model is to decide the purpose of the model in terms of the predictions desired and to define the time and space boundaries to be used in the model. Time boundaries define the time period over which we want the predictions to apply. This is important because it will determine whether short- or long-term processes need be considered in the model. Spatial boundaries specify whether the model will predict conditions in a pond, a hectare, a watershed, 1000 hectares, or the Province of New Brunswick; these boundaries will determine whether or not we need to incorporate processes such as emigration/immigration, variability in weather conditions over distance, etc.

Another area of problem specification that can be useful is establishing subsystems for the model. Establishing subsystems (and thus submodels) can be very important, particularly in complex models; subsystems are more easily modeled and examined; submodels can be used independently or not; and submodels reduce the likelihood of programming errors by reducing program length. Criteria for choosing subsystems are numerous and are discussed by Goodall (1974). In general, a good guideline for establishing subsystems is given in Munn (1975) as "... smaller areas of the problem which, although highly interconnected internally, have relatively few links with other parts of the system." This aspect can be considered at these initial stages and also later on, after flow charts are constructed.

At this specification stage, simple flow diagrams are often helpful. These flow diagrams indicate the parts of the system and their possible interconnections. Note that these flow diagrams are not those used later to specify the detailed sequence of events in the model (see Section III,C,2).

Finally, we may further specify the problem by making a list of the possible (and/or available) variables and their interrelationships. This list begins to indicate what data are available and what data may be relevant, although unavailable at this time. However, our list of variables at this time may well have unnecessary entries and/or lack some important entries, both of which may be detected in steps that follow.

To bound the fiddler crab problem, we will consider the population of fiddler crabs inhabiting the borders of creeks and channels in a couple of hectares of the salt marsh studied in New Jersey. By restricting the model application to creek and channel borders, where the crab populations are densest, we can eliminate, at least initially, consideration of movements between high density and low density areas (this movement was determined to be rather minimal, at least during the period spanned by data collection). As to time boundaries, we can limit model application to a 5-year period. This time specification requires us to include reproductive processes in the model, since 5 years will span a number (probably about 10) of generations. However, the period is short enough to ignore genetic changes in the population, at least initially. In terms of subsystem specification, we will not construct subsystems, since this model is extremely brief and simple for illustrative purposes. As previously noted, however, the present model could itself be a submodel of an overall marsh ecosystem model.

Simplified flow diagrams are often helpful in these initial specification stages to clarify the parts of and the connections among parts of the system. Thus we can represent the fiddler crab system as in Fig. 3.1. While constructing the flow diagram, several decisions were made. First, insecticide effects will be considered only in terms of mortality, at least initially. Any sublethal effects that may affect the population will not be considered (we will see later that such sublethal

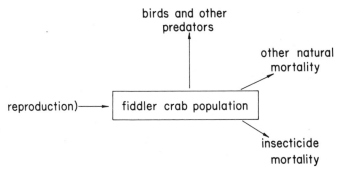

Fig. 3.1.   Major factors affecting the numbers of fiddler crabs on the marsh.

effects are important in this case). Another simplifying assumption made was that birds and other predators do not consume dead fiddler crabs; this would be of some importance if we were considering changes in bird and other predator populations. Lastly, we assume that fiddler crab food is never in short supply, since marsh soils are generally rich in organic matter and bacterial and fungal films. We now have a very simplified diagram of the fiddler crab system, but one that will be useful as a modeling illustration. We can convert our diagram to a tautology as follows:

Change in No. of fiddler crabs/m² = (increase due to reproduction)−(loss due to predation)
− (loss due to other "natural" mortality)−(loss due
to insecticide mortality)

We can now consider possible variables to be used in the model. One list of such variables could be the following:

1. Number of fiddler crabs per unit area
2. Number of offspring per crab per unit time (reproduction rate)
3. Percent of population being lost to predators/time
4. Predator density
5. Percent of population dying from "natural causes"
6. Percent mortality due to insecticide applications
7. Insecticide application rate and frequency
8. Effect of physical factors (tidal washing, temperature) on insecticide at time of application

This initial listing begins to suggest some important questions. For example, since it seems necessary to include fiddler crab predators (known to be primarily clapper rails, willets, and dowitchers but perhaps also including some fish, blue crabs, and some sparrow species), will we also need to construct a population model for each predator species? In this case, we will decide against this alternative, since we want to keep the model as simple as possible and since basic information on the birds and other predator species is not abundant. Therefore, we can represent the predation on fiddler crabs as a rate varying with the density of fiddler crabs; that is, we assume a constant relatively high and homogeneous population of predators. Another way of dealing with decisions of this kind is described in Section III,D. For the present, however, the above list is suggestive of data and relationships we may use in constructing the model. In general, it is not wise to draw up an initial variables list in great detail or length, since the steps that follow will be much more helpful in selecting appropriate variables.

B. Choosing and Specifying the Relationships of Interest: Interaction Table and Functional Relationships

Having considered the basic questions and possible variables (factors) for the model, we need to identify factors that interact with other factors. An interaction table is of help in this task. Such a table cross-lists the factors we are considering for inclusion in the model. For example, an interaction table for the fiddler crab problem may be as shown in Table 3.1. The boxes marked with an X in Table 3.1 indicate a direct effect of the factor listed in that row on the factor entered in that column. Note that in a normal model there would probably be more interactions; the scarcity of interactions here results from the extremely simplifying assumptions made for illustrative purposes.

Another general use of interaction tables, applicable in problems of greater complexity (more compartments or subsystems) than the fid-

TABLE 3.1.

Interaction Table for Fiddler Crab Problem

	Effect on				
Effect of	Reproduction rate	Predation rate	Natural mortality rate	Insecticide mortality rate	No. of crabs
Reproduction rate					X
Predation rate					X
Natural mortality rate					X
Insecticide mortality rate					X
No. of fiddler crabs	X	X			X

dler crab example, is to generate the variables that best represent the interactions between subsystems. For example, if we were dealing with a model of a fiddler crab population, a predator (bird) population, and a detrital food source for the crabs, we might lay out an interaction table as shown in Table 3.2. The object is to enter, in the appropriate boxes, the variable(s) of the "from" subsystem that we need to know to simulate what happens to the "to" subsystem. For example, what do we need to know about the detritus subsystem to model the effects on the fiddler crabs? It is clear that we do *not* need to know *everything* about detritus dynamics, etc., for this purpose. Thus, one important advantage of this method is to avoid the trap of collecting masses of unnecessary information about all compartments. What we do need to know about detritus is those factors whose inclusion in the model will significantly alter the predictions of the model. Such factors are variables directly affecting the density or dynamics of fiddler crab populations, such as organic matter content and size of the

TABLE 3.2.

"From" or "effect of"	"To" or "effect on"		
	Detritus	Fiddler crabs	Birds
Detritus		Organic matter content Size of particles	
Fiddler	Density of crabs Consumption rate Defecation rate Organic matter content of feces		Density of crabs Catchability of crabs
Birds	Density of birds Organic content of feces and defecation rate	Density of birds Use of alternate food sources Food selection behavior (size, etc.) Daily food requirements	

particles. These latter variables determine the energy intake of the fiddler crab population, which is directly related to population increase. Similarly, other variables can be entered in the table as shown. This procedure is much more effective in selecting appropriate variables than simply pondering what variables may be needed to construct a model. Most often, making an interaction table results in changes in the initial "possible variables" list. The table has the effect of focusing on the *interactions* between subsystems and away from the subsystems themselves.

Once the interaction variables have been generated, we can identify additional variables (to be entered in the detritus–detritus, fiddler crabs–fiddler crabs, and birds–birds boxes) that will be needed to provide interaction variables. For example, in the fiddler crab–fiddler crab box, we need to know reproductive rate and how crab density affects reproductive rate to calculate a new value for crab density, which is needed for the detritus and bird subsystems.

As a result of the preceding steps we now have a list of variables we will use to describe the state of the system to be modeled. This list of state variables is called a system state vector. In addition, we have specified some interaction relationships between state variables.

Finally, and most importantly, we want to specify the functional relationships that govern changes in the system; that is, we want to define the *form* of the interactions between state variables. To do this, we examine one variable at a time and state how the variable will change. Variables we examine may be specified to remain constant, to vary as a function of other variables, to be composed of several variables (each of which may vary independently), etc. The rules for change of a variable may be expressed as a simple equation (linear, simple-form curvilinear) or as a graph whose equation is not specified. Graphs may be entered directly into computer models, or a mathematician can provide the equations with the appropriate graphical shapes. Watt (1968, p. 265) lists the equations for a number of common types of relationships.

To specify a functional relationship, the change in the variable must be expressed as a function of different states of the system; that is, we should measure the change in variable X as Y changes in value, rather than the change in X with time. Then we may have a graph as shown in Fig. 3.2. This graph expresses a functional relationship between X and Y. Note that in many field situations such data will be unobtainable under normal conditions, since a range of states of Y may not occur. In these cases, the measures at different states of Y can be

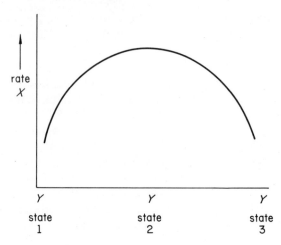

Fig. 3.2. A functional relationship between X and Y.

obtained by manipulation of the field conditions or by laboratory experiments, or the graph can be assumed to follow theoretically predicted shapes. It is in this sense that mathematical modeling is useful in directing research; it points out relationships that are needed to understand (and simulate) the overall functioning of the system but that may not be empirically available. Holling (1961, 1965) presents an approach based on the interplay between systems modeling and experimentation that he calls experimental components analysis. Often, we will need to literally invent (with the aid of some known facts about the variables in question) a graph to express a functional relationship that is important (i.e., that seems likely to determine, to some appreciable extent, the results of the model's predictions). For an example of formulating such a relationship, see the reproduction of fiddler crabs used as an illustration below. The ability to invent functional relationships allows us to (1) proceed with the modeling effort, (2) explore the sensitivity (the dependence of the model's predictions or outcome) of the model to this relationship, and (3) to decide whether or not to gather experimental data on that relationship, depending on the results of the initial model. As preparation for sensitivity analysis, it is a good idea, at this stage of functional relationship definition, to specify and have a written record of all reasonable alternative hypotheses for each interaction and to make approximate estimates of maxima, minima, and thresholds.

It is extremely important at all times in modeling to keep in mind that the predictions of the model are strictly dependent on these assumed relationships, and as such the predictions may or may not differ drastically from the actual system's behavior.

To formulate functional relationships for the fiddler crab model, we examine the list of state variables and the interaction table given previously. We need to express the density of fiddler crabs as a state variable, say UCA (the genus name of the crabs). It is advisable to always keep an updated written list of all variable names, since confusion often arises when the model attains some complexity and/or when some time passes since the variables were named. The next factor we need to express is the reproductive rate of the crabs. We note from the interaction table that we must express the interaction between number of crabs and reproductive rate in some form. We have no data on the form of this relationship in fiddler crabs, but we guess from studies on many other species that we can probably expect density-dependent reproduction (see brief review in Watt, 1968, pp. 288–311). We can, therefore, invent and scale a reasonable functional relationship between reproductive rate and crab density. An Allee-type curve, which seems the most general, is of the form shown in Fig. 3.3. There is an optimal, intermediate population density where we can expect the highest reproductive rates. At high population den-

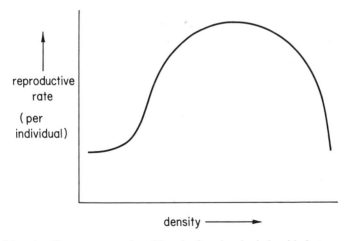

Fig. 3.3. An Allee-type curve describing the functional relationship between reproductive rate and density.

sities, interference and competition phenomena reduce reproductive rates. This can be expected in fiddler crabs, since they are territorial and aggressive and since they live in permanent underground burrows. At very low population densities, the Allee curve also predicts a reduced reproductive rate; such effects may be due to factors such as difficulty in locating mates below a threshold level of population density. The simplest (although by no means the only) way to represent this curve for making estimates in the model is to redraw it as a series of straight-line sections as shown in Fig. 3.4. This graph can be easily entered into a computer model as a series of straight-line functions, each one of which is brought into use when population density values are between the appropriate limits. The specification of these functions will be shown below once we have scaled the graph for the fiddler crabs.

Having chosen a shape for the functional relationship, we can estimate the values of density and reproductive rate for which the graph may apply. Ullyett's (in Watt, 1968, p. 293) data for reproduction of moths shows maximum reproductive rates between one-half and two-thirds of the maximum population density value, and minimum reproductive rates below one-sixth of the maximum population density. Experimentally observed maximum population density for fiddler crabs was about 175/m² (Ward *et al.*, 1976); therefore, we estimate 200/m² as a possible maximum density value, at which reproductive rate may be quite low. Note that this value can be changed for sensitivity analysis of the model (see Section III,D). Using the relative popula-

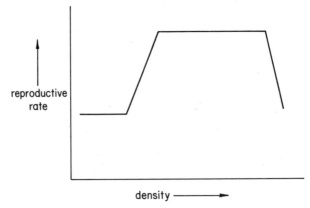

Fig. 3.4.   A simple portrayal of the functional relationship shown in Fig. 3.3.

tion densities quoted for Ullyett's moths, the fiddler crab graph would be scaled as shown in Fig. 3.5. We will, however, make one adjustment, in the lower density area of the curve: since *Uca* can certainly find each other at a density of 20/m², the threshold density for increased reproductive density when scaled according to the moth data, we will set the threshold for very low reproductive rates at a density of 5 crabs/m², and correspondingly expand the range of densities over which maximum reproduction occurs. We now need to scale the reproductive rate axis. From field observations, it appears that the population of *Uca* under study maintained a fairly constant density from year to year. Since winter survival must be relatively low (say at most 50% survival) we can assume that the population doubles during the summer. Thus, the reproductive rate axis can be scaled as shown in Fig. 3.6. Once again, we can vary the maximum reproductive rate for sensitivity analysis; if the model's predictions turn out to be very sensitive to this factor, we may decide that research on this factor should be conducted.

Now we have a fully scaled functional relationship, for reproductive rate versus population density, that can be used in the model. The equations to be entered for calculations are as follows.

Population density value (number/m²)	Equation for reproductive rate (RR = number of young/adult-season)
0–5	RR = 0.1
5–77	RR = 0.0056(UCA)+0.072
77–128	RR = 1.0
128–200	RR = −0.0056(UCA)+1.211

These equations are easily obtainable from the scaled graph. It should be reemphasized that all of these assumptions can be manipulated, during sensitivity analysis, to see how much difference they make to the model's predictions.

Returning to the list of factors we are specifying and to the interaction table, we see that we must formulate the loss of fiddler crabs due to predation and its interaction with density of fiddler crabs. Again, we have no data on this interaction, but we may represent it by a Michaelis–Menten-type graph derived from Holling's (1959, 1965, 1966) theoretical work on predation processes. Such a graph is shown

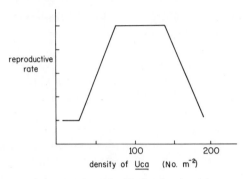

Fig. 3.5. The graph shown in Fig. 3.4 with the density axis approximately scaled.

in Fig. 3.7. This relationship allows for increasing predation as the prey abundance increases, until the maximum feeding rate possible is reached at very high prey densities. The shape of this graph can be described by the equation

$$\text{Feeding rate} = \frac{A \text{ (prey density)}}{(B + \text{prey density})},$$

where A is the maximum (saturation) feeding rate, B is the prey density at which the feeding rate is half of the maximum (see previous graph), and the feeding rate is the number of prey eaten per individual predator per unit time. The parameters A and B set the shape of the curve. From studies on a variety of species, B is often 40% of the density value at A; we can, therefore, start by assuming this ratio

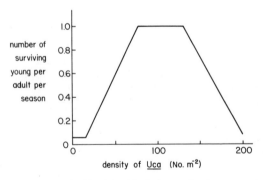

Fig. 3.6. The graph shown in Fig. 3.5 with the reproductive rate axis scaled as well.

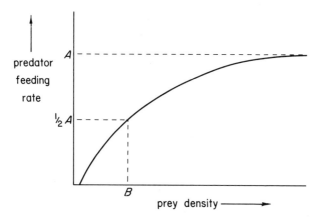

Fig. 3.7. A possible functional relationship between predator feeding rate and fiddler crab density.

between B and the density value at A. [For a more realistic application of this predation model, see Eggers (1975).]

To scale the feeding rate–prey density functional relationship for fiddler crabs and their predators, we will again make some rather arbitrary assumptions, but assumptions that can be explored and tested empirically later. We will assume that the maximum number of fiddler crabs that a clapper rail can eat in one day is 10 (a guess based on the relative body sizes of the two species) and that this feeding rate can be attained at a fiddler crab density of 100/m². The equation for the fiddler crabs is

$$\text{Number of } Uca \text{ eaten per rail per week (UPRW)} = \frac{A(\text{UCA})}{(B + \text{UCA})},$$

and we will initially let $A = 70$, $B = 40$ Uca/m² (UCA = number of fiddler crabs/m²). These parameters will be varied for sensitivity analysis. The loss of fiddler crabs due to predation can be obtained by the product of UPRW and the density of rails per square meter. The density of rails per square meter will be varied in the model, but we may want to make an initial estimate. Based on the data of Ferrigno (1966), a likely figure for clapper rail density in the summer in this marsh might be 0.01 to 0.1/m². We now have specified another functional relationship for use in the model.

We have now represented two common biological processes, reproduction and predation, as functional relationships. These relationships will introduce feedback dynamics into the model by reason of the form of the functions. We could formulate functional relationships for some of the remaining variables, but we are attempting to keep the model very simple. Therefore, we will treat the rest of the factors as driving variables (variables that are not themselves calculated by the model). Thus, we will represent death from "natural causes" other than predation and mortality due to insecticide use as flat rates that can be input as different values. In addition, we will not consider the effects of physical factors on the insecticide, since field observations and experiments showed that the crabs ingested the insecticide directly and quickly after application on the marsh surface (Ward *et al.*, 1976). We have specified (at least for the "first round") ways of representing each term of our basic model equation:

Change in number of crabs/m^2 = (increase due to reproduction) − (loss due to predation) − (loss due to other "natural mortality") − (loss due to insecticide mortality)

The system equations are

Increase due to reproduction = (UCA) (0.1)   *OR* (UCA) (0.0056 × UCA + 0.072)   etc.

Loss due to predation = (UPRW × RAILS)
where UPRW = $\dfrac{(A \times \text{UCA})}{(B + \text{UCA})}$

Loss due to other natural mortality = UCA × DINAT
where DINAT equals a flat rate of mortality such as 0.10

Loss due to insecticide mortality = UCA × RMRABT
where RMRABT equals a flat rate of mortality due to Abate, such as 0.20.

C. Programming

To this point in the analysis of the fiddler crab–insecticide use problem, we have systematically bounded the questions for consideration, chosen the relationships and links likely to be influential in the outcome of our predictions, and formulated ways of representing those relationships. This systematic analysis is of value in any research endeavor, whether or not a computerized model is to be used. As a result of the analysis, we have a precise and careful description that was chosen after sorting through a variety of ways of representing the system. We have identified specific areas for research that would contribute directly to prediction of selected outcomes, and ruled out other areas that, although perhaps interesting for other reasons, are not directly important for making the relevant predictions. Thus, the explicit modeling process is of considerable value even if a computerized full-scale model is not constructed.

1. *Criteria for Computerization*

The next step is to decide whether a computer model would be of value. Criteria for making this decision are outlined in Munn (1975); here I will list and briefly discuss some of those criteria. A computer-based mathematical model is generally of help if (1) a large volume of data (i.e., of simple calculations) needs to be handled, (2) the conceptual model contains many complex links between its elements, (3) it is important to determine changes over time as a result of the proposed actions, (4) increased definition of underlying assumptions and components of the system appears desirable, (5) some or all of the relationships between the components of the system can only be expressed in terms of probabilities rather than exact values, (6) there is a good possibility of defining the essential components of the system and the relationships between the components (the two remaining criteria given in SCOPE 5 are not relevant to our discussion of biological system analysis, but relate to modeling entire socioeconomic–biological systems rapidly, and without experimental follow-up studies, for policy analysis).

a. *Volume of Data.*   In describing almost any ecological system, no matter how simple, at least several interacting components are included. Each of these components will be represented in terms of a list of state variables, each of which may assume a number of values. Thus, in even very simplified systems, such as the fiddler crab model discussed here, a considerable volume of data exists. The relationships specified in the model will require calculations to be performed on those data, so that we quickly amass large numbers of simple calculations that need to be executed rapidly. Computers are ideal for rapid performance of large numbers of simple calculations, as well as for systematic storage of resulting values to be used in future calculations.

b. *Complexity of System Relationships.*   Most ecological models, including the fiddler crab model, have a number of links between elements. If we attempt to mentally visualize changes in the system as the various relationships are operating, we quickly realize that we cannot comprehend or keep track of even a limited number of links operating together, nor of indirect effects that may result. Further, some of the relationships are nonlinear; nonlinear influences are particularly difficult to assess mentally. Computerized mathematical models are well suited to handle many relationships at once as well as to explore the effects of indirect and nonlinear relationships.

c. *Consideration of Rates of Change.*   Many ecological processes occur at varying rates, and environmental changes affecting those processes may also occur or be introduced at varying rates. Delays and time thresholds may also exist in the processes being modeled. If the processes being modeled appear to contain such dynamic aspects, and exploration of the rates of change seems important, a computer-based model will be helpful.

d. *Explicit Relationships.*   The explicit nature of mathematical models was discussed as a major advantage in Section I. Definition of relationships forces recognition of underlying assumptions and points out areas where information may be insufficient. Computerization of a mathematical model forces even more precise definition than explicit conceptual models, since actual computational formulas and sequences are specified.

e. *Uncertainty.*   In ecological studies many elements and relationships will not be known with certainty, but will have to be estimated as a range of probable values or a series of possible relationships. We have seen that estimation of this sort is valuable in that it allows us to proceed with the modeling effort which is of value in identifying crucial knowledge gaps and assumptions. The variability in the input (as values or relationships) can be used by the model to indicate the probable range of the effect or outcome. This feature is called stochastic (as opposed to deterministic) modeling. As mentioned previously, uncertainty can be handled by exploring the model's response using various inputs; this procedure is also important in performing sensitivity analysis (see Section III,D). The repeated calculations involved can be performed easily with computers.

f. *Knowledge of the System.*   Although some input to models can be educated guesswork and probability functions, we must have a basic knowledge of the system being modeled in order to construct the model. The usual sources of this rough information are literature survey and field surveys. Enough knowledge must be available to allow definition of basic components and relationships between components. Although this criterion applies initially to conceptual modeling, it should also be considered when deciding on computerization.

However, it is important not to apply this criterion to the point of spending years gathering data before a model is constructed. Most often, enough information about a system is available from rapid literature search and field observation to construct a model in a short time and before much or any experimental information is gathered.

2. Model Implementation

a. *Flow Charting.* Once the functional relationships have been developed, as outlined in Section III,B, it is useful to construct a flow chart of calculations or bookkeeping system for applying the relationships in a calculation sequence. Many systems for flow charting have been used, but the simplest forms are the best, particularly for non-programmers. An example is given in Fig. 3.8 for the fiddler crab model. The sequence of steps specified by the chart is intended to represent the real time sequence of events that is to be simulated by the calculations. Although we are limited somewhat by the step-by-step nature of a computer's calculations, and thus cannot simulate real processes in their simultaneous operation, short time intervals between calculations simulate most phenomena quite adequately.

b. *Programming.* Once the basic relationships have been specified and the flow chart constructed, programming is relatively simple. It consists of translating the flow chart into a series of statements readable by computers. FORTRAN is the most widely used computer language, and numerous manuals and short courses are available for rapid learning of FORTRAN. Other languages are also available, as well as a variety of languages that have been created expressly for simulation (DeWit and Goudriaan, 1974; Radford, 1971). Simulation languages allow use of computers without much knowledge of programming, but by the same token the modeler does not have a good grasp of the operation of the model on the computer. Another tack is to employ a programmer to implement the model on the computer while consulting with the modeler-biologist. Lastly, there are also simulation control systems (using FORTRAN or other basic languages) that control input and output of the computer program and allow changes to be made while the model is running (Hilborn, 1973).

Using flow charts, the FORTRAN statements can be written step-by-step until we come to the last statement in the flow chart. An example, again for the fiddler crab problem, is shown in Fig. 3.9. This model was programmed in FORTRAN IV for the IBM 360/67 computer system at the University of British Columbia and uses the simulation control

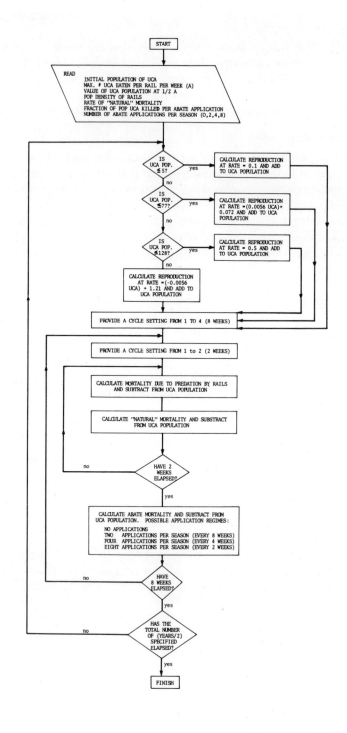

system SIMCON (Hilborn, 1973). It is useful to work systematically, defining one step at a time, to keep a carefully up-dated list of variable names being used, and to keep programming as simple as possible, at least in the beginning. Consultation with programmers can help in many cases.

Some model output for the fiddler crab example is presented in Fig. 3.10. The predicted *Uca* populations under several insecticide regimes are shown; under this set of assumptions, the *Uca* populations can stand at most four applications of 5 lb/acre Abate per season, each causing 5% mortality or two applications of the same Abate per season, each causing 10% mortality.

D. Sensitivity Analysis

Having run the model on the computer, the modeler searches for possible inconsistencies or unrealistic results by deliberately changing some of the input, either as assumptions or variable values, and studying the effects on the model output. Generally, assumptions, variables, or relationships that do not seem to influence model outcome can be ignored or dropped from the model, thus simplifying it. Those assumptions, variables, or relationships that seem very influential to model outcome should be pursued further in simulations. Often, this kind of analysis will lead to selection of areas for experimental research into aspects for which the model has great sensitivity; such aspects sometimes turn out to be key components in the biological

---

Fig. 3.8. Flow chart for fiddler crab computer model. Rectangles usually represent calculation steps, while diamonds represent branching decision steps. The chart specifies that the program is to read the state variables list first. Then it is to calculate the value of the crab population density after reproduction (the rate of reproduction applied depends on the population density at the time of the calculation, and reproductive increase is calculated twice each year). Then the program is to provide cycle settings so that mortalities due to insecticide applications may be calculated at intervals of 8 weeks, 4 weeks, or 2 weeks during a 16-week summer season, and so that other mortality can be applied every 2 weeks over the summer (we assume that the crabs are torpid and in their burrows for 8 months each year; this period varies considerably with climate). The program must then calculate new crab population density values after application of rates of predation and other natural mortality. Then the program is to calculate new crab density values after insecticide mortality is applied according to a chosen schedule for this run. Then the program is to stop the calculations when the specified number of cycles is reached.

```
#LIST CRAB2.S@-IC
>    1              SUBROUTINE UINIT
>    1.25 $CONTINUE WITH CRAB2.C RETURN
>    2              CALL CMREAD ('CRAB2.C ')
>    3              CALL DFAULT ('1=CRAB2.D ')
>    4              RETURN
>    5              END
>    6              SUBROUTINE UMODEL (IYEAR)
>    7      $CONTINUE WITH CRAB2.C RETURN
>    8      C
>    9      C CALCULATE POP UCA AFTER REPRODUCTION
>   10      C
>   11              IF (UCA.LE.5.) GO TO 10
>   12              IF (UCA.LE.77.) GO TO 20
>   13              IF (UCA.LE.128.) GO TO 30
>   14              GO TO 40
>   15           10 UCA=UCA+(UCA*0.1)
>   16              GO TO 50
>   17           20 UCA=UCA+(((.0056*UCA)+.072)*UCA)
>   18              GO TO 50
>   19           30 UCA=UCA+(UCA*0.5)
>   20              GO TO 50
>   21           40 UCA=UCA+(((-.0056*UCA)+1.211)*UCA)
>   22           50 CONTINUE
>   23      C
>   24      C ENTER SUMMER LOOP
>   25      C
>   26              DO 1000 J=1,4
>   27      C
>   28      C ENTER TWO-WEEK LOOP
>   29      C
>   30              DO 500 I=1,2
>   31      C
>   32      C CALCULATE MORTALITY DUE TO PREDATION BY RAILS
>   33      C
>   34              UPRW=A*UCA/(B+UCA)
>   35              UCA=UCA-(UPRW*RAILS)
>   36      C
>   37      C CALCULATE NATURAL MORTALITY
>   38      C
>   39              UCA=UCA-(UCA*DINAT)
>   40          500 CONTINUE
>   41      C
>   42      C CALCULATE ABATE MORTALITY
>   43      C
>   44              IF (ABTIND.EQ.0) GO TO 1000
>   45              IF (ABTIND.EQ.1) GO TO 990
>   46              IF (ABTIND.EQ.J) GO TO 990
>   47              IF ((ABTIND*2).EQ.J) GO TO 990
>   48              GO TO 1000
>   50          990 UCA=UCA-(RMRABT*UCA)
>   51         1000 CONTINUE
>   52              RETURN
>   53              END
#END OF FILE
#LIST CRAB2.D+CRAB2.C
>    1      S #IDUMP=ON
>    2      S UCA=40.
>    3      S A=70.
>    4      S B=70.
>    5      S RAILS=.01
>    6      S DINAT=.005
>    7      S RMRABT=.15
>    9      S ABTIND=0.
>    1              COMMON UCA, DINAT, RMRABT, ABTIND
>    2              COMMON A, B, RAILS, UPRW
#END OF FILE
```

system. Although our fiddler crab model is too simple to present a realistic sensitivity analysis, other examples are deWit and Goudriaan's (1974) model of growth of a corn fungus, Diamond's (1974) model of cycling of potassium in the Hubbard Brook forest, and Walters and Peterman's (1974) systems analysis of spruce budworm in New Brunswick (see Chapter 6).

E. Model Verification and Validation

Model verification consists of running the simulation model, as in sensitivity analysis, with a range of inputs and comparing the output with theoretically expected behavior of the system. Gross inconsistencies will generally reveal obvious errors in underlying assumptions or variable and relationship selection. Model modification then often produces a model that seems to capture the behavior of the system to the best of our understanding.

Model validation is, in contrast to verification, a comparison of the model with reality. Model validation becomes particularly important when model results are to be used for decision making about the system's resources. In validation, model behavior must be compared to a new set of data, distinct from those used to construct the model. Such an independent set of data may be obtained by (1) holding back a part of the available data when first constructing the model or (2) collecting more experimental data for model validation. Collecting special data for model validation, although not always feasible in environmental impact assessment, has considerable advantages. First, data collection can be very selective, concentrating on parameters in common with model output. Second, having the guidance of the hypothetical model, several kinds of model predictions can be tested, even if some of these predictions are not important to impact assessment. If the model is validated by comparing several of its predicted results with new data, our confidence in its mimicry of real processes is greatly increased.

Fig. 3.9. Computer program, in FORTRAN IV and using the simulation control system SIMCON, for the fiddler crab model flow-charted in Fig. 3.8. The statements follow in the order specified by the flow chart. The first seven statements are SIMCON control statements. Descriptions of the calculation-specifying statements are preceded by the letter "C." The first nine lines of the second file listed ("# LIST CRAB 2.D + CRAB 2.C") is the system state vector, in which we may enter any desired beginning values for the variables to be used in the program.

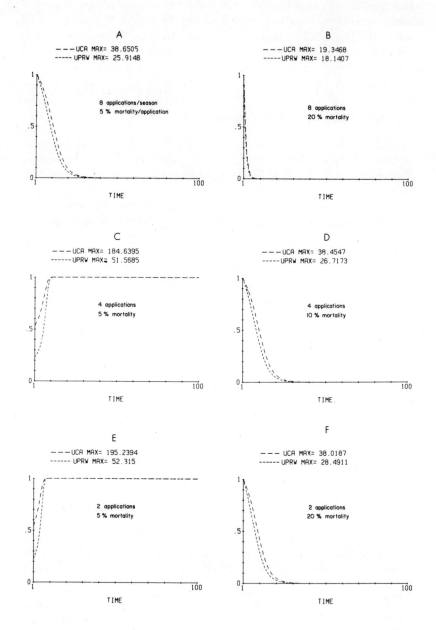

Fig. 3.10. Some output of the fiddler crab model. The y axes show the density of *Uca* per square meter (long dashed lines) plotted as a fraction of the maximum density

Holling (1978) convincingly argues that model validation is in reality a process of invalidation in that it is designed to show where the model fails. In particular, a useful way to examine where a model fails is to attempt to predict qualitatively different behaviors observed in ecological systems similar to the one we have modeled. If we can mimic these behaviors using the model with small plausible changes in parameter values or model structure, our belief in the model's future predictive power increases. If we cannot mimic these behaviors, degree of belief in alternative models may be increased.

REFERENCES

Clark, W. C., Jones, D. D., and Holling, C. S. (1978) *Ecol. Model.* In press.
deWit, C. T., and Goudriaan, J. (1974). "Simulation of Ecological Processes." Cent. Agric. Publ. Doc., Wageningen.
Diamond, P. (1974). *Proc. Int. Congr. Ecol., 1st, The Hague* pp. 16–21. Cent. Agric. Publ. Doc., Wageningen.
Eggers, D. M. (1975). A synthesis of feeding behavior and growth of juvenile sockeye salmon in the limnetic environment. Ph.D. Thesis, Univ. of Washington, Seattle.
Ferrigno, F. (1966). Some aspects of the nesting biology, population dynamics, and habitat associations of the clapper rail. M.S. Thesis, Rutgers Univ., New Brunswick, New Jersey.
Goodall, D. W. (1974). *Proc. Int. Congr. Ecol., 1st, The Hague* pp. 244–249. Cent. Agric. Publ. Doc., Wageningen.
Hilborn, R. (1973). *Simulation* **20**, 172–175.
Holling, C. S. (1959). *Can. Entomol.* **91**, 385–398.
Holling, C. S. (1961). *Annu. Rev. Entomol.* **6**, 163–182.
Holling, C. S. (1965). *Mem. Entomol. Soc. Can.* No. 45.
Holling, C. S. (1966). *Mem. Entomol. Soc. Can.* No. 48, 1–85.
Holling, C. S., ed. (1978). "Adaptive Environmental Assessment and Management." Wiley, New York.

---

attained during that simulation (that maximum is given as UCA MAX with each graph), and the predation rate by rails (UPRW=number of *Uca* eaten per rail per week), also plotted as (short dashed lines) a fraction of the maximum attained during that simulation run. The x axes are time in years, each simulation having been run for 100 years. Note that although the model's time boundaries had been set at about 5 years, such that running the model for 100 years is not justified, most of the changes occur in appropriately short time frames. Each graph shows the effects caused by the stated number of applications, each application effecting the mortality stated. These predictions are strictly dependent on the model's assumptions and are presented for methodological illustration only.

Munn, R. E., ed. (1975). "Environmental Impact Assessment: Principles and Proce-
dures," SCOPE (Scientific Committee on Problems of the Environment), Rep. No.
5. Int. Counc. Sci. Unions, Toronto.
Patten, B. C., ed. (1971). "Systems Analysis and Simulation in Ecology." Academic
Press, New York.
Peppard, L. E. (1975). *Int. J. Syst. Sci.* **6**, 983–999.
Pielou, E. C. (1972). *Science* **177**, 981–982.
Poole, R. W. (1974). "An Introduction to Quantitative Ecology." McGraw-Hill, New
York.
Radford, P. J. (1971). *In* "Mathematical Models in Ecology" (J. N. R. Jeffers, ed.), pp.
277–295. Blackwell, Oxford.
Walters, C. J., and Efford, I. E. (1972). *Oecologia* **11**, 33–44.
Walters, C. J., and Peterman, R. M. (1974). *Quaest. Entomol.* **10**, 177–186.
Walters, C. J., Hilborn, R., and Peterman, R. (1975). *Ecol. Model.* **1**, 303–315.
Ward, D. V., Howes, B. L., and Ludwig, D. F. (1976). *Mar. Biol.* **35**, 119–126.
Watt, K. E. F., ed., (1966). "Systems Analysis in Ecology." Academic Press, New York.
Watt, K. E. F. (1968). "Ecology and Resource Management." McGraw-Hill, New York.

4

The Field Experiment

I. INTRODUCTION

Two major approaches to predicting development impacts on ecosystems exist. In one approach, we study the structure and function of the natural ecosystem and, based on that knowledge, formulate hypotheses on likely effects of development actions. This approach rests on the assumption that some understanding of the nature and functioning of an ecosystem will allow predictions of responses of that system to manipulations. Thus, many ecological impact assessments are made on the bases of a literature survey, checklist and matrix impact identification methods, and/or a field survey. A second approach assumes that in order to know the effects a development will have we must proceed with the development actions in some form (simulated development actions on experimental plots, monitoring of similar developments in other areas, etc). Experimental manipulations of the ecosystem and/or monitoring of similar actual developments are not used commonly in ecological impact assessment, since such work requires considerable resources and time. However, some assessments involve extensive survey work that could be more profitably converted, at least in part, to experimental and monitoring studies. One great advantage of manipulative studies is that we see states of the system that may not have occurred in the past and that may involve different functional responses than could be expected on the basis of study of the unimpacted system.

II. EXPERIMENTAL SYSTEMS

There are a variety of ways in which experimental information about effects of development actions can be obtained. Some experimental systems are purely research devices, in which manipulations are performed on a system and responses are measured. Other systems combine experimentation with real development actions and management programs in various ways. Possible field experimental systems of both kinds are as follows:

1. Subject a portion of the system in question to simulated proposed actions
2. Subject a smaller, comparable system to simulated proposed actions
3. Monitor ongoing or completed development action effects
4. Allow a scaled-down selected version of the proposed development to proceed; based on the resulting observed impacts, allow further development increments in modified versions (adaptive management)

In the following sections each of the above field experimental systems will be considered.

It should be noted that each of the above types of experimental studies can be performed using different approaches and at many levels of detail; the questions of which variables to measure, how many variables to measure, and what level of replication to use are always important. One method of providing a focus is to study any changes in those variables or processes or compartments that appear to play major roles in ecological function of the system, such as productivity, decomposition, key species, etc. (see Chapter 7, Section V). Another method for focusing on critical variables and interactions is mathematical modeling as described in Chapter 3. Modeling and, particularly, sensitivity analysis as a guide to experimental studies can be profitably used in combination with all the experimental system approaches listed above. This method can reduce monitoring and experimental system designs to manageable proportions in many cases. The question of level of replication in field experiments is discussed in Section IV,D.

A. Subject a Portion of the System to Simulated Development Actions

This experimental approach can be very useful in predicting development action impacts, since many of these impacts will occur in the experimental situation. It is very applicable to the study of effects of additions of substances such as pesticides, heavy metals, sewage, etc. As in one of the examples discussed in Chapter 6, experimental plots can be set up and the substance to be added can be used on the experimental plots; a variety of effects may be measured as a result. Depending on the system being affected, this approach can also be applicable to study of effects of changes in temperature, salinity, water regime, physical features of relatively small scales, and density of organisms (generally only in organisms with little or no mobility or in certain territorial organisms). This approach is particularly useful when the system to be affected is relatively uniform and large. For example, a forest, a grassland, a marsh, a large lake, or a coastal intertidal zone often contain enough area of similar features to provide enough comparable plots for statistical treatment of the resulting data. These conditions may not be met by smaller or more heterogeneous systems, such as a stream or a river.

Important limitations of this experimental scheme include spatial scale and time scale effects. Spatial effects were already briefly mentioned above. In some cases, applying the development action to a small area may not show effects that would emerge if the whole system were affected, since dilutions and movements of organisms in and out of the test plots may reduce or otherwise alter the impacts of the experimental treatments. In addition, this approach may not be applicable to certain large construction developments, such as building dams and jetties. Time scale effects must also be considered, since experimental plots are usually only studied for a few years at the very outside, and often for only one summer, while the effects of the real development actions will generally occur for a much longer time. Thus, effects that involve processes with time lags (including many basic biological processes, such as reproduction and predation effects on populations) and effects that are cumulative in nature may not be detected in relatively short-term field studies, whether they be of this type or of other types. It should be noted, however, that information from short-term field studies of this type, when combined with other short-term approaches can provide indications of longer-term effects that can then be pursued individually. For example, in the study of effects of an organophosphorous insecticide (temephos) on salt marshes in New Jersey, short-term field data from experimentally treated plots showed that the abundant marsh snail *Melampus bidentatus* retained measurable amounts of temephos for more than 5 weeks after the last treatment with a granular formulation of the compound (Fitzpatrick and Sutherland, 1976). This raises the question of whether, given longer-term exposures to this treatment, animals feeding on *Melampus* containing temephos would accumulate the compound to toxic levels. Similarly, any evidence of changed behaviors due to short-term field exposures to a disturbance would provide indications of possible population effects to be expected and studied further.

B. Subject a Small Comparable System to Proposed Actions

In some cases, a part of the system cannot be employed as an experimental unit, either because of the heterogeneity and relatively small size of the system or because effects on the whole system are expected to differ drastically from the effects observable in component parts of the system. In such cases it may be valuable to consider the use of a whole smaller system for experimental purposes. For example, a stream may be used as a model of a river; the whole stream can be experimentally manipulated and monitored for resulting changes. Extreme caution must be used in extrapolating the results, however, since scale effects can be very important. A well-known example is the absence of stratification in small lakes as opposed to large lakes. Similar but less obvious differences may make some extrapolations from small systems to larger, apparently comparable, systems totally unfounded.

The similarities and differences of this experimental approach to the use of physical model systems are worth mentioning. Physical model systems, or microcosms, are discussed in Chapter 5. Both approaches use a model system for experimentation, but one is a natural system while the other is created for the experiment. While the natural model system offers more complexity and realism, there is usually less control over random fluctuation. Both types of model systems present problems in extrapolating to the larger system being modeled. The artificial model system may offer the slight advantage of being more explicitly only a partial model of the real system.

C. Monitor Ongoing or Completed Development Effects

For many kinds of developments monitoring studies on previous similar developments in other areas help to forecast possible effects. Caution must be used in extrapolating results directly, however, since an apparently similar site may differ markedly in chemical and physical characteristics and in biological phenomena. Even the same or similar species may behave differently under different climatic regimes. Thus site-specific monitoring studies are important even when considerable information on development impact is available.

Monitoring studies can include a predevelopment and a post-development phase. Most often the preoperational study period is very limited or nonexistent. Predevelopment monitoring periods are important because they provide one kind of control condition, allowing before and after comparisons. Ideally, a side-by-side comparison also should be included in a monitoring study. Often, comparable unaffected nearby sites are not available for this purpose (see Chapter 1, Section IV,B). Postdevelopment monitoring is important because many effects will involve time lags of considerable duration. Long-term monitoring is valuable because such long-term effects can be detected and because the information obtained can be used to suggest development modifications to ameliorate the unforeseen long-term effects.

It is important to explore a variety of approaches to a monitoring study, as was stressed for impact studies in general in Chapter 1. Too often, monitoring studies have focused too narrowly on components of the ecosystem expected to show some effect and on the individual mortalities of those species. It is of value to consider possible system effects and population effects of any mortalities or other individual changes. For example, monitoring studies of large developments, such as power plants (steam and nuclear), have been numerous and have tended to take a narrow focus on immediate mortalities of selected species. Coutant (1969) and Ichthyological Associates (1972) review a variety of such studies. To predict the impact of such mortalities or sublethal effects, population models of the species in question are needed. This is particularly important when there is a possibility of cumulative effects occurring. For example, small fish mortalities due to thermal effects of one power plant may become of major importance to the fish populations if a number of power plants are placed along the same coast. Similarly, without population models it is impossible to judge the population effects of factors such as increased fishing success in warm water areas around power plants, where fish may congregate in the cooler seasons of the year. Population models of such situations may suggest whether overexploitation of stocks is likely to result in such situations. At least population modeling focuses attention on such possible effects that have been ignored in the narrow studies.

D. Adaptive Management

Adaptive management is not strictly an experimental approach to ecological impact study, but it can produce data that can be used for impact assessment. Adaptive management (Holling, 1978; Walters and Hilborn, 1976) uses the initial stages of a development to gather information about possible later effects of that development and formulates changed development plans for further development. The changed development plans have a dual purpose: to minimize adverse effects (or maximize desirable effects) and to yield further information about the system's responses. The dual purpose entails trade offs between present system performance and knowledge of system potentials and disaster thresholds. Therefore, in some cases suboptimal (for whatever set of characteristics is being optimized) management will be used purposely to gain information. In this sense, adaptive management constitutes an experimental approach to ecological impact studies. Experiments of this sort can be very simple, involving a pilot study in which a small portion of a proposed development is studied for resulting effects, or quite complex, using formal mathematical methods to decide on policy or development changes. An example of the latter is Walters and Hilborn's (1976) analysis for exploitation rates of fishing systems. These authors suggest that fisheries managers should occasionally underfish and overfish to test their estimates of the productivity of the fish population.

Adaptive management seeks to extend the range of the system's responses known to the manager. One advantage of such knowledge is reduction of possible unexpected consequences. Operating at a single productive or "stable" point on a system's response space reduces our capability to predict departures from such points.

In the case of large construction developments, economic and population growth considerations sometimes dictate a plan of development to be accomplished in several stages. Monitoring and experimental studies done in conjunction with the development stages can, therefore, form an adaptive management unit. In other cases, developments involving substantial unknown factors (such as fishery enhancement programs) can be broken down into small units that can be manipulated separately for maximum information yield, rather than to apply a large homogeneous management program to a large geographical area. Thus, adaptive management strategies can be used to exploit a trial and error approach to environmental impact. Some of the impacts discovered in this manner cannot be predicted by studying the intact system and on that basis making guesses as to possible development impacts.

III. TIME AS A FACTOR

In Chapter 1, Section IV,D, some emphasis was placed on the notion that simultaneous use of a variety of experimental techniques often provides indications of ecological changes to be expected in widely differing time frames. In this section, time frames for ecological impacts are discussed further, and methods most suited to deal with each class of time frames are suggested.

Short-term ecological impacts of development or use actions may be loosely conceptualized as those changes taking place in time frames of at most weeks or months. Examples are acute toxicity, physical changes in habitat (barriers, current changes, etc.), immediate reproduction, predation and competition effects in short-lived species (plankton species, micro-organisms), some behavioral changes, and in some cases seasonal effects. Short-term changes are generally the easiest to detect and measure. Methods such as direct experimental manipulation, short-term sampling, and laboratory experiments are very applicable to the study of short-term impact.

Medium-term phenomena occur in time frames of several years. Such effects can include changes in macrophyte productivity and changes in populations of larger organisms due to disturbances in reproduction, predation, competition, etc. The most applicable methods in this time frame are generally combinations of simulation modeling and experimental work in the field and the laboratory.

Long-term phenomena can be thought of as occurring in time frames of decades or longer. Effects commonly occuring in this time frame are significant genetic changes in large organisms, climatic changes, food web magnification of substances, and many cumulative effects of incremental development or actions (e.g., buildup of sufficient air pollution sources in a region to affect plant physiology, etc.). The methods most suitable to study and predict such long-term effects are modeling, extrapolation from present development demands, and long-term monitoring programs.

It should be clear that the above classification of time frames is arbitrary and is only intended to focus attention on the need to consider possible ecological impacts occurring in widely differing time frames. It is also clear that many phenomena, including eutrophication, succession, and food web magnification may occur at quite different rates depending on factors such as the rate of pollutant addition, size of the system affected, etc. Often, immediate indications of these impacts can be detected, but the process continues into considerably longer time frames. The above examples of effects to be expected in the various time frames generally occur in those time spans, but can vary in some specific situations.

IV. SOME SAMPLING PROBLEMS

A. Sampling Objectives

In ecological impact studies sampling may be performed to accomplish a variety of objectives, either as part of a preliminary field survey or as a means of hypothesis testing in designed experimental studies of the types described in Section II. As part of a preliminary field survey, sampling may serve to provide the following: a qualitative description of the species present, information on some species characteristics (such as feeding relationships, dominance, and certain behavioral patterns), and preliminary information on species distribution and possible relationships of species distribution to environmental variables. Two very important points should be made in this regard. First, it should be noted that many, if not most, environmental impact studies consist of only this preliminary, field survey type of information and do not include experimental hypothesis-testing field studies. This is a major weakness in most environmental impact studies. Second, it is often the case that field surveys are done too extensively. That is, the objective of only getting enough data to allow a rough qualitative description is often lost sight of, and survey sampling becomes too extensive to also permit experimental manipulative work. In hypothesis-testing experimental studies, more intensive and selective sampling may be performed to estimate the following quantitatively: selected measures of species abundance or productivity, degree and nature of predation and competition among species, behavioral changes, and patterns of species distribution and relationships of those patterns to various parameters.

It is of great importance to specify clearly the objectives of any sampling effort before the work begins, since the nature of the objective will determine the type and amount of sampling to be done. For qualitative field survey purposes, statistical considerations are not important, and most sampling procedures will be adjusted to spot sampling along the physical gradients or in the distinct areas that are evident in the system being sampled. It is usually wasteful and of little utility to sample intensively at this stage before the following planning steps have been taken: (1) assemble literature and field survey information into a qualitative model of the system, (2) identify processes and variables likely to be crucial to ecosystem function and/or to model outcome (see Chapters 2 and 3), and (3) design formal hypothesis-testing experiments relating those processes and variables to manipulated factors (environmental variables, development actions, etc.). In this context, the term "design" of experiments includes plans for the field lay-out, the apparatus to be used, hypotheses to be tested, and the specific statistical treatment to be given to the data gathered. The statistical treatment of the data will partly specify the sampling strategy (see Section IV,D). The importance of these planning steps cannot be overemphazied.

B. Absolute and Relative Measures

Having decided what variables to measure in an experimental or monitoring field study, we must decide what measures of those variables to use. Absolute measures, such as total populations, total nutrients in pools, etc., are usually impractical and often not informative. Relative measures, such as population per square meter, milligrams of nutrients per liter, etc., are usually used. In many cases, serious biases or errors may be known to exist in even relative measures. For example, the litter bag method of estimating decomposition rate (see Chapter 2, Section III,D) probably involves serious errors of estimate. One reason is that the material in the bags is subject to quite different flow of the medium from the natural loose material. Thus, not only is mechanical breakup probably decreased in the bags, but also rate of wash-out of breakdown products is reduced. Nevertheless, standardized procedures in litter bag studies allow their valid use to compare decomposition rates in treated versus control plots. Similarly, the relative abundance indices discussed in Chapter 2, Section II,A can be very useful to compare different experimental conditions rather than investing the excessive time and resources needed to measure population density accurately in some cases. The disadvantage of many relative measures, of course, is that the data gathered cannot be used for comparisons with other systems and other sampling procedures.

C. Delimiting Sampling Areas

Considerations for delimiting sampling areas depend heavily on the particular characteristics of the system to be sampled. For example, aquatic systems generally pose very different problems from those present in terrestrial systems. Nevertheless, there are a number of basic considerations that apply to setting up sampling areas in most kinds of systems.

The environmental impact study usually specifies the natural system to be considered for ecological investigations. The first step is to conceptually divide the system to be studied into its component ecosystem types. For example, we may divide an estuarine area into bay, main river channel, marsh areas, bank areas, etc. A somewhat simpler system, a forest, may be divided into an oak–hickory area, a beech–maple area, and so on. In some cases, the system of interest may include many different ecosystem types, necessitating selection of at most a few of these for detailed study.

In each type of ecosystem, preliminary observation will indicate possible sampling areas. The size of the area to be sampled will vary with many factors. The following considerations will put approximate size requirements on the area:

1. Habitat heterogeneity: If an ecosystem to be sampled appears fairly heterogeneous, it will probably be necessary to do stratified sampling (see Section IV,D) in the area. Stratified sampling will usually require a larger total area than would be needed to do random sampling in a relatively homogeneous area.

2. Experimental design: If certain hypotheses are to be tested by the sampling program, a number of matched plots or areas will be necessary to provide for control plots and different test conditions. Thus, enough area must be allowed to set up comparison plots, each encompassing similar major physical features, and for any spatial separation necessary among plots (for example, spatial separation to prevent contamination of control plots.)

3. Species mobility: The mobility of species of interest will determine whether relatively small plots or areas can be subjected to experimental manipulations or whether large geographical areas will be needed as test plots and study of movements to and from the test plots will be required. For example, study of most invertebrate benthos, terrestrial infauna, and many territorial organisms can be carried out in relatively small treated (manipulated) test areas. On the other hand, study of very mobile organisms, such as flying insects, birds, and many nonterritorial vertebrates, cannot generally be done in experimental plots of a few hectares or less, since the organisms are sure to move in and out of the test plots.

4. Species abundance: Obviously, if the population(s) being sampled is dense, a smaller area will be required for sampling than if the species has small numbers per unit area.

5. Type and size of sampling gear: The type of sampling gear will vary greatly depending on the physical medium and the taxonomic group being sampled. The size of gear, such as quadrats, transects, and nets, will also vary with these factors. Both type and size of gear will have an influence on the size of the area to be set aside for sampling. It should be noted that sampling need not always involve removal of a part of the population from the experimental plots. In many cases observational counts or remote sensing of many types can be used for sampling.

D. Sampling Strategies

Selection of a sampling strategy depends partly on the objectives for sampling. For example, whether a measure of population density or a measure of component factors, such as natality, mortality, and migration, is desired may specify the choice between a quadrat (removal) sampling method or a mark–recapture sampling program. In addition, the heterogeneity or homogeneity of the system to be sampled affects the choice of sampling strategies. A variety of previously used sampling strategies exist and should be considered before a sampling program is initiated. Table 4.1 presents some of the major strategies available and some of their characteristics and applicability. The table should be consulted in the discussion that follows. Figure 4.1 shows a flow diagram representing one possible scheme for selecting a sampling strategy. This diagram will also be referred to throughout the discussion that follows.

The objectives for sampling may determine the sampling strategy as shown in Fig. 4.1. Detailed population dynamics information is obtained by methods such as mark–recapture and change in composition of stocks (Poole, 1974; Watt, 1968) applied over the long term. Estimates of population abundance may be obtained by a variety of direct sampling methods that can be applied with random, systematic, stratified, or cluster sampling strategies. Simple classification of population abundance levels into a few classes can be performed with sequential sampling. The latter strategy is especially applicable when a large area or region must be sampled regularly (Poole, 1974). It should be noted that although the flow chart in Fig. 4.1 and this discussion will explore in detail only the sampling strategies to estimate population abundances, similar considerations of heterogeneity and homogeneity of the system will apply when planning strategy for factors such as trap placement in terrestrial mark–recapture studies.

Having decided on a direct sampling method for the field system and having delimited the tentative sampling area, the next step in choosing a sampling strategy is to examine the field system for the presence of any distinct parts or strata that differ from the other parts. If apparent strata exist, the sampling area can be divided into those strata and preliminary sampling in each area can be performed for an analysis of variance. This ANOVA is a check for significant heterogeneity in the sample area. The tentative strata are used as the treatments of the ANOVA. Basic statistics textbooks, such as Snedecor and Cochran (1967), Schor (1968), or Sokal and Rohlf (1969), describe the procedures involved in ANOVA. Morris (1955) contains an example of a rather complex three-way ANOVA performed for this purpose. If ANOVA shows significant effects of these tentative strata, then stratified sampling should be used. On the other hand, if the area does not show significant heterogeneity, one of the strategies useful for homogeneous systems is applicable (see Table 4.1). As noted in Fig. 4.1, this analysis detects fairly obvious heterogeneity in that we observe apparently different areas and as-

TABLE 4.1.
Major Sampling Strategies

Sampling	Sampling area or organisms homogeneous or heterogeneous	Special features	References
Stratified	Heterogeneous	Sample allocation important	Poole (1974); Cochran (1963); Watt (1968)
Random (includes transects)	Homogeneous	Sample size determination	Poole (1974); Cochran (1963)
Two-stage	Homogeneous		Poole (1974); Watt (1968)
Systematic (aligned random, unaligned Latin squares)	Homogeneous or heterogeneous	For convenient sampling of homogeneous area or for testing for nonobvious heterogeneity (caution needed to avoid introducing systematic bias)	Poole (1974); Cochran (1963); Yates (1948, 1960)
Cluster	Heterogeneous in clusters	Cluster "grain" must be taken into account in size of sampling unit	Cochran (1963)
Sequential	Heterogeneous or homogeneous	Characterizes sample into population classes	Morris (1960); Poole (1974)
Mark-recapture (also survey removal, change in age composition of stock, etc.)	Heterogeneous or homogeneous	Estimates components of rate of population density change	Krebs (1972); Poole (1974); Watt (1968)

Sampling objective: estimate

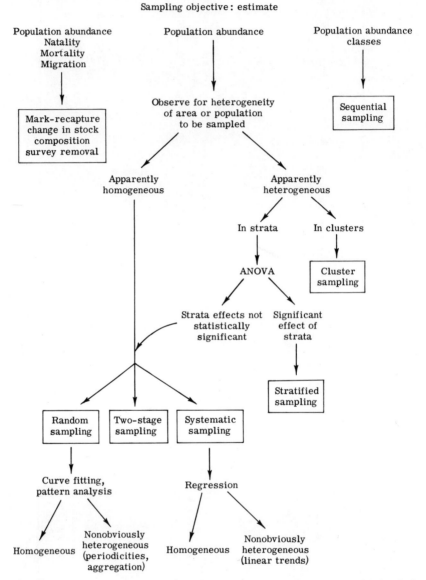

Fig. 4.1.   A flow diagram for selection of sampling strategies. See text for detailed discussion.

sign those areas to be treatments in the ANOVA. Less obvious heterogeneity may be detected by performing regressions on other measured variables (for methodology, see Poole, 1974; Watt, 1968). However, the preliminary data gathered for this initial check on heterogeneity are sometimes not suitable for regression. If nonobvious heterogeneity is suspected, it can be examined in more detail after a systematic sampling strategy (see below) designed with the intention of doing regressions as well as estimating population abundance.

The initial observations to detect heterogeneity may indicate a clumped or clustered distribution of the organisms to be sampled. The clusters may be quite evident, in which case clusters to be sampled are selected at random, and sampling is performed in those selected clusters. Cochran (1963) discusses techniques and problems of cluster sampling. Poole (1974) reviews methods for detecting less obvious cluster distribution and other dispersion patterns and includes methods for detecting cluster size.

If the initial ANOVA shows significant effects of the strata established observationally, stratified sampling can be performed. In stratified sampling the usual procedure is to sample independently and at random in each stratum. Allocation of equal numbers of samples can be made to each stratum; this procedure is convenient and appropriate for many biological situations, especially at the initial stages of sampling. If more information about the strata is available and if cost of sampling is a consideration, optimum sample allocation may be determined. In optimum sample allocation, those strata that are more variable are sampled more intensively and those strata in which sampling is more costly are sampled less. Cochran (1963) discusses optimum sample allocation in detail; Poole (1974) presents a more concise discussion of the same topic.

Referring again to Fig. 4.1, let us suppose that our initial observations did not reveal any obvious heterogeneity in the area and/or population to be sampled. In this situation, we may apply random sampling, two-stage sampling, or systematic sampling.

Random sampling can be used in a homogeneous area to obtain an unbiased estimate of the mean population density. A random numbers table or some other source of random numbers is used to select the individuals or quadrats or sample locations to be sampled from all possible samples. The principle underlying random sampling is that if samples are selected at random we are unlikely to cause, by our sample selection, a systematic deviation of the estimated means from the true population mean. A major consideration in random sampling is determining the sample size to be used.

An appropriate sample size for random sampling can be derived in a number of ways. One method, described in various forms by many statistics textbooks, is based on a theoretical statistical argument. We begin by assuming that we are dealing with a normal sampling distribution of the mean, with mean $\mu_{\bar{X}} = \mu$ and standard deviation $\sigma_{\bar{X}} = \sigma/n^{\frac{1}{2}}$ (where $\mu_{\bar{X}}$ is the mean of sample means, μ is the population mean, $\sigma_{\bar{X}}$ is the standard deviation of sample means, σ is the population standard deviation, and n is the sample size). In such a distribution, 95% of the sample means ($\bar{X}$) will fall in the interval (approximately) $\mu_{\bar{X}} \pm 2\sigma_{\bar{X}}$, and 5% of the means will fall outside that interval. If we can specify, for a specific sampling situation, a desired confidence interval (I) about the true population mean (μ), we can state that an acceptable probability of the sample mean ($\bar{X}$) falling outside this range ($\mu \pm I$) is, for example, 5%. This confidence interval can then be expressed as $I = 2\sigma_{\bar{X}}$, which can be estimated by $I \simeq 2s_{\bar{X}}$ ($\sigma_{\bar{X}}$ is the standard deviation of sample means, while $s_{\bar{X}}$ is its estimate). Since $s_{\bar{X}} = s/n^{\frac{1}{2}}$ (s is the sample standard deviation), $I = 2s/n^{\frac{1}{2}}$. Squaring both sides we obtain $I^2 = 4s^2/n$ or $n = 4s^2/I^2$. This n is the minimum necessary sample size to ensure this confidence interval for 95% of sample means.

If analysis of variance is to be carried out on the sampling data, the ANOVA requirements can also be considered as an aid to determining sample size. Poole (1974) contains an excellent concise discussion of ANOVA that can be used for review purposes. Sampling replication is necessary in ANOVA if it is desired to test for the presence of statistically significant interactions between factors. Since interactions are common in biological systems and sampling errors are often large, considerable replication is required. As a rough guideline, at least five replicates in each condition are desirable to estimate error variance in ANOVA.

Another way of arriving at an appropriate sample size for random sampling is to examine pilot data and determine a suitable sample size empirically. For example, Fig. 4.2 shows a graph of the average standing crop of roots and leaves of two species of marsh grass (*Spartina alterniflora* and *Spartina patens*) versus number of samples of standing crop taken. The total series of samples consisted of ten replicates, since this was an absolute upper limit on replication set by practical constraints. The idea was to examine the gain in stability of the mean

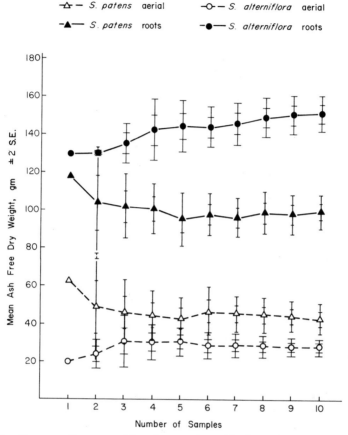

Fig. 4.2.   Mean standing crop of *Spartina* as estimated by increasing sample sizes.

and error as sample size was increased from 1 to 10. Figure 4.3 shows a graph of the standard deviation of the mean standing crop (one measure of spread) as sample size increases. By looking at these graphs, it was decided that the increment in stability of the measures of mean and variance with more than five replicates was insufficient to warrant greater replication. That is, after $n = 5$, fluctuations of the mean and standard deviation seem considerably reduced compared to those observed at $n < 5$.

A last method of arriving at an appropriate sample size is based on experience in sampling in a variety of biological systems. In most

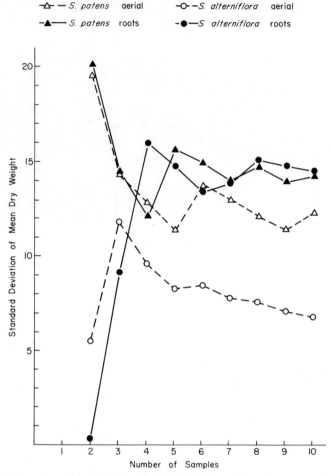

Fig. 4.3. Standard deviation of mean standing crop of *Spartina* as sample size increases from 1 to 10. Note that after $n=4$ or $n=5$ there is relatively little gain in stability.

biological sampling in the field involving a number of test conditions, ten or more replicates are impossible for practical reasons. Thus an approximate upper limit for replication is nine or ten samples. A lower limit is set by the need to estimate error variance. One or two samples is useless for this purpose; three or four may be marginally acceptable under some circumstances; and five is probably the usual lowest limit. Thus, we are restricted, in most cases, to a range of five to nine or ten replicates. If you can, take eight to ten; if that is impossible, take as many as possible, but at least five.

Two-stage sampling is also applicable for homogeneous areas. The area is divided into smaller units (primary units) and then a random sample of the smaller units is taken. The primary units chosen at random are subsampled; that is, a random sample of individuals is taken in each chosen primary unit. Poole (1974) gives calculational methods for the number of primary and secondary units to be used for certain estimated standard errors of the estimate and for varying costs of sampling. It should be noted that two-stage sampling is equivalent to one-stage random sampling when all the primary units are subsampled. Which of the two strategies is more advantageous depends on the cost of sampling and the variability between and within primary sampling units for a specific sampling program.

Systematic sampling may be used also in homogeneous situations, that is, when the organisms being sampled are distributed essentially at random. It is often convenient to take systematic samples, and the precision in these situations is approximately equivalent to that obtained by a stratified random sampling strategy (Cochran, 1963). In many cases, however, the system being sampled is only apparently homogeneous and contains hidden linear trends or hidden periodicities. Hidden linear trends can often be discovered by aligned systematic sampling and regression against measured environmental variables. Unsuspected periodicities, however, can introduce serious biases in the estimates made by systematic sampling, unless special precautions are taken. For example, systematic samples taken in a field may be arranged as shown in Fig. 4.4. If the object is to estimate the population mean by systematic sampling with minimal bias, the unaligned design is far superior to the aligned scheme (Cochran, 1963). These unaligned systematic sampling designs have been termed Latin squares. Cochran (1963) and Yates (1960) discuss details of Latin square designs. It should be stressed that the unaligned pattern shown in Fig. 4.4 represents a systematic intentional lack of alignment. A random pattern would in most cases contain more alignment than the systematic unaligned design.

Now, what if an area or population distribution is apparently spatially homogeneous but contains hidden periodicities, nonlinear trends, or aggregation patterns, and we wish to discern those spatial dispersal patterns rather than make unbiased estimates of population means? The usual approach in this case is to use random sampling

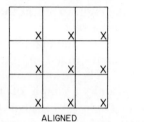 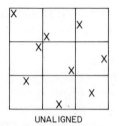

ALIGNED   UNALIGNED

Fig. 4.4.   Aligned versus unaligned systematic sampling. Note that the unaligned design purposely excludes any alignment of samples.

and a variety of data analysis techniques, such as iterative regression (Watt, 1968), fitting of common distributions, and pattern analysis methods (Poole, 1974).

It should be emphasized that the above discussion of sampling theory is an attempt to provide a framework for considering a variety of available sampling strategies as possibilities for sampling in an ecological impact study. It does not intend to thoroughly discuss any one specific sampling technique. Such discussions can be found in the references cited in Table 4.1. The intent of this discussion is to provide applicability guidelines so that a choice of sampling strategies can be made more efficiently and in awareness of the major possibilities.

E. Experimenter-Induced Bias

One kind of experimenter-induced bias, that introduced by the way in which a particular person takes a sample as opposed to a different person, has already been discussed as an example in Chapter 1, Section IV,B, where it was pointed out that counterbalancing sampling schemes can be used to combat this source of bias.

Another type of experimenter-induced bias is due to the fact that when an experimenter removes samples from an ecosystem, he or she is acting as a consumer of the organisms or materials sampled (Falk, 1974). Further, if these samples are removed and never returned to the environment, as is usually the case, the experimenter acts as a consumer with 100% efficiency, since no calories or elements return to the environment from the material consumed. Thus, every field sampling program presents the problem of evaluation of the ecological significance of that sampling under those particular circumstances. Falk (1976) suggests a rough technique for estimating such impact, based on literature values of utilization efficiency, pilot data estimates of experimenter consumption and population levels, and comparison of experimenter calorie consumption to calorie consumption by other consumers in the system. Such sampling consumption can be minimized by changes in the sampling procedure (especially as to timing of sampling in populations with rapid turnover) and by reducing sample size. In Section IV,D the determination of sample size was discussed primarily as a function of statistical significance and secondarily as a function of sampling cost. Now a third factor to be considered in determining sample size, especially when estimating natural consumption, mortality, or energy flow, appears to be experimenter-generated consumption levels for the proposed sample size for a particular situation. At present, this factor can only be estimated roughly but can be evaluated on that basis. For example, Falk (1974) estimated that Odum's (1957) energetic study of Silver Springs, Florida, the experimenter's total caloric consumption was equivalent to one-third of that of the entire population of the largest major carnivore. This study probably represents an extreme example of experimenter consumption on a relative basis, since there was only an estimated population of three of these largest major carnivore (a gar) and the system sampled was small and isolated from potential sources of rapid replacements for sampled organisms. However, even when studying larger less isolated systems from the points of view of population dynamics, competition, and energy flow, the impact of sampling consumption should be considered and at least be reported as an estimated bias.

Another aspect of experimenter-induced bias, even when sampling does not constitute consumption of the organisms or materials being sampled, is the effect of the experimenter's activities on the behavior of the organisms being sampled. In some cases, this impact is quite obvious and can sometimes be counteracted by performing similar actions on control plots of some sort. For example, in the study of effects of insecticide use on salt marsh ecology described in Chapter 6, one treatment consisted of spraying temephos-bearing clay granules from a helicopter, at 2-week intervals. The effects of the helicopters' operation on the behavior of the birds and of all animals on the plots was not known. However, this effect was removed as a source of variance when comparing treated plots with control plots by the simple device of flying the helicopter, in an identical manner to that used in the treated plots, over the control plots each time a treatment was applied. These flights over control plots involved no treatment at all or applied insecticide-free blank granules. This latter treatment was made to check on possible effects of the clay-like granules on the system.

In other cases, behavioral effects elicited by the experimenter's presence and activities may be more difficult to control. One commonly encountered example is the effect of having trapped an animal in a mark–recapture study. Often animals become "trap happy" (having been previously trapped changes their behavior so they are more likely to be trapped again than animals not previously trapped) or "trap shy" (previous trapping makes an animal less likely to be recaptured than its naive fellows). Either of these two experimenter-induced behavioral changes violates a basic assumption of the mark-recapture technique. In this method, all animals in the population are assumed to be equally likely to be trapped. Cormack (1966, 1969) discusses approaches for handling these problems.

It is important to consider possible effects of experimenter-induced bias carefully in designing any experiment. In field experiments, experimental impacts may take the form of adding a source of consumption of organisms and of competition with other consumers or of induced behavioral changes in the animals present. Experimental designs should account for any such suspected impacts. In some cases, approximate quantification can be made, as suggested by Falk, and if the impact appears significant, adjustments can be made in the sampling procedure or extent. In other cases, experimental controls for the experimenter's impact can be designed so that control areas are subject to the same impacts as treated areas.

V. USE OF RESULTS

The results of field experiments such as those described in this chapter can, in some cases, be used directly to formulate predictions of expected impacts of proposed development actions. For example, if a field experiment is performed using simulated development actions and an ANOVA design, we can evaluate the measured effects of the simulated development factors and their interactions as to their statistical significance. This evaluation provides a basis for prediction of real development action effects. This approach is more or less a purely empirical approach.

Another main use of experimental field study results is to support, correct, or revise our models of the system to be affected by development and of likely effects of development. The interaction of field experimental studies (including the resource management-experimental system hybrids) and modeling is very powerful, and allows more rapid predictions of impact and exploration of uncertainties in the system (see Chapter 3). Thus, in this use field experimental results provide input for biological submodels, that in turn are input for overall impact assessment models (see Fig. 1.1).

REFERENCES

Cochran, W. G. (1963). "Sampling Techniques," 2nd Ed. Wiley, New York.

Cormack, R. M. (1966). *Biometrics* **22**, 330–342.

Cormack, R. M. (1969). *Annu. Rev. Oceanogr. Mar. Biol.* **6**, 455–506.

Coutant, C. C. (1969). "Thermal Pollution—Biological Effects (A Review of the Literature of 1969)," Rep. No. BNWL-SA 3255. Battelle Mem. Inst., Pacific Northwest Lab., Richland, Washington.

Falk, J. H. (1974). *Oikos* **25**, 374–378.

Falk, J. H. (1976). *Ecology* **57**, 141–150.

Fitzpatrick, G., and Sutherland, D. J. (1976). *Pestic. Monit. J.* **10**, 4–6.

Holling, C. S., ed. (1978). "Adaptive Environmental Assessment and Management." Wiley, New York.

Ichthylogical Associates (1972). "Ecological Considerations for Ocean Sites off New Jersey for Proposed Nuclear Generating Stations." Ithaca, New York.

Krebs, C. J. (1972). "Ecology." Harper, New York.

Morris, R. F. (1955). *Can. J. Zool.* **33**, 225–294.

Morris, R. F. (1960). *Annu. Rev. Entomol.* **5**, 243–264.

Odum, H. T. (1957). *Ecol. Monogr.* **27**, 55–112.

Poole, R. W. (1974). "An Introduction to Quantitative Ecology." McGraw-Hill, New York.

Schor, S. (1968). "Fundamentals of Biostatistics." Putnam, New York.

Snedecor, G. W., and Cochran, W. G. (1967). "Statistical Methods," 6th Ed. Iowa State Univ. Press, Ames.

Sokal, R. R., and Rohlf, F. J. (1969). "Biometry: The Principles and Practice of Statistics in Biological Research." Freeman, San Francisco, California.

Walters, C. J., and Hilborn, R. (1976). *J. Fish. Res. Board Can.* **33**(1), 145–159.

Watt, K. E. F. (1968). "Ecology and Resource Management." McGraw-Hill, New York.

Yates, F. (1948). *Philos. Trans. R. Soc. London, Ser. A* **241**, 345–377.

Yates, F. (1960). "Sampling Methods for Censuses and Surveys," 3rd Ed. Griffin, London.

5

Laboratory Studies

I. INTRODUCTION

Previous use of laboratory studies to predict effects of anthropogenic actions has consisted primarily of simple experiments in which organisms collected in the field or reared in the laboratory are subjected to a stress and the amount of mortality and/or the presence of abnormalities are studied. A common form of output of such studies is a series of dose–response curves for the organisms selected. Unfortunately, the use of this approach in isolation from more complex laboratory studies, from field studies, and from explicit modeling has led to two important problems: (1) the results of the laboratory studies may not extrapolate to field conditions and (2) a proliferation of irrelevant laboratory studies may result. Extrapolation to field conditions of results of isolated laboratory studies is often dangerous or impossible, since interactions of laboratory-observed effects with population and community interactions, with physicochemical fluctuations, and with other field conditions is often dramatic. The problem of extrapolation of laboratory study results is discussed in detail in Section II. The second major problem arising from use of isolated laboratory experiments for ecological impact predictions has been one of inefficiency, since little information is available to select critical aspects to be studied in detail. Thus, a large number of laboratory experiments may be performed, and many may turn out to be totally irrelevant to the outcome of the predictions being made while the critical experiments may be omitted. This latter point has already been made with respect to field experiments. Explicit system modeling can be of great help in selecting specific aspects for empirical investigation (Chapter 3, Section III,B).

Productive efficient use of laboratory experimentation for ecological impact predictions involves laboratory work as one interacting element of a number of approaches. It is a means of performing relatively rapid controlled hypothesis testing when guided by field work and explicit modeling and when the complexity of the laboratory test system is carefully considered. Often, test system complexity can be varied to allow us to extrapolate from laboratory results with greater confidence.

In general, laboratory studies that are coupled with other approaches and whose complexity is adequate can be profitably used for two purposes in impact studies. One use is direct testing of effects of a stressing factor on the biological system (this chapter). The other general use is for determining the form and the scale of some functional relationships in interaction with explicit modeling (discussed primarily in Chapter 3).

II. EXTRAPOLATION FROM LABORATORY STUDIES

The basic problem involved in the process of extrapolation is that of identifying factors present under field conditions that are likely to alter or eliminate the effects observed in the laboratory. In what follows I will list and discuss some general classes of field factors previously found to alter stress effects observed under laboratory conditions.

A. Changed Nature of the Stressing Factor When Applied under Field Conditions

Certain stresses may be transformed under field conditions such that impact on a species or system differs markedly from that expected from laboratory experiments. An example is the compound DDT and a host of other organic pesticides. DDT may affect a species directly or may degrade to DDE and DDD. The factors determining the form of the compound present include environmental conditions and metabolic processing by organisms; thus, the form predominating will vary under different conditions. The toxicity of these three compounds generally differs considerably for any one species, so that it is important to determine which forms of the compound are present, their relative amounts under the field conditions, and the physiological conditions in question. An illustrative case is the finding that the compound associated with eggshell thinning in birds is DDE rather than its parent compound DDT (see studies reviewed in Pimentel, 1971; Stickel, 1973). Another relevant example is the biological (and chemical) conversion of inorganic mercury, which chemists believed would be fairly inert in the environment, to methylmercury compounds. The methylated mercury compounds are more highly toxic and more biologically mobile than other forms of mercury (D'Itri, 1973).

B. Interaction of Stress Effects with Variability in Physicochemical Factors

Given that a pollutant or other stress factor has an impact on an organism under optimum conditions of temperature, salinity, etc., it has been commonly found that the stress effect is magnified and sometimes becomes lethal under sub-optimal physicochemical environmental conditions. For example, Vernberg and Vernberg (1972) showed that fiddler crabs exposed to mercury concentrations that were sublethal under optimal salinity and temperature suffered significant mortality in a short time under stressful temperature and salinity conditions. Thus, the effect of mercury on fiddler crabs becomes more severe under expected environmental conditions.

As an example of a stress effect that may become less severe under conditions to be expected in the field, Gillott et al. (1975) showed that the presence of sediment in test systems was highly significant in reducing photosynthetic inhibition in the alga *Euglena gracilis* by relatively low concentrations of three different insecticides. Since the sorptive capacity of the sediments for the insecticides was the factor accounting for the reduced toxicity to the algae, areas containing sediments of differing coarseness and organic matter content would depress the insecticide toxicity to differing degrees. Thus, simple laboratory exposures to insecticides are not appropriate to predict field toxicity to the algae.

C. Cumulative Effects of Stress with Time

Some stress effects accumulate with time; thus, short–term laboratory experiments can severely underestimate expected field effects. Accumulation of effects generally involves underlying accumulation of a substance in organisms. Accumulation of substances may involve a single amassing or cumulative storage in an organism or other site or it may involve food web magnification of persistent substances (Woodwell, 1967). By either of these mechanisms, a time period would be involved for the accumulation to occur. Further, by the food web magnification mechanism an extensive food web must be present to detect the phenomenon. Therefore, it is unlikely that simple short-term laboratory experiments can predict such cumulative stress effects.

D. Individual Characteristics of the Tested Species under Field Conditions

Under field conditions, a species may show behavior not displayed in the laboratory; it may be present at a number of different developmental or life stages, not all of which are tested in the laboratory, and it may be found in (sometimes drastically) different physiological states from those occurring under laboratory conditions. All of these differences in the state of the species between laboratory and field conditions often lead to field impacts that differ markedly from those expected on the basis of laboratory results. For example, in the field a fish may show an avoidance reaction to a thermal effluent and thus not suffer expected elevated temperature effects. The adult stage of a crab may not show mortality when exposed to certain concentrations of mercury, but larval stages may be quickly killed by the same concentrations (De Coursey and Vernberg, 1972). The state of nutrition of a bird may well affect or determine whether elimination of one of its usual feeding habitats will prove lethal.

E. Intra- and Interspecific Interactions

Population and community interactions, such as competition, predation, and reproductive success or failure, may interact with anthropogenic stresses to produce population and/or community changes quite different from those expected from dose–response data. Such ecological interactions may affect either lethal or sublethal (changes in behavior, metabolic rate, reproduction, etc.) effects of a stress factor. An example of such an interaction with a lethal effect of stress is Ryther's (1954) study of photoplankton and oysters mentioned in Chapter 2. The addition of organic nutrients to two bays in Long Island proved lethal to the normal dominant phytoplankton community of diatoms, green flagellates, and dinoflagellates and favored the growth of two previously unimportant small green flagellates. The oysters and other shellfish, which had previously utilized the normal dominant phytoplankton, were unable to digest the new dominants and died out in the bays.

An example of community interactions affecting sublethal stress effects is the study of organophosphorous insecticide effects on marsh fiddler crabs in New Jersey (Ward et al., 1976; Ward and Busch, 1976). Marsh plots that were caged over to exclude major predators of fiddler crabs did not show significant changes from control plots in fiddler crab population density when treated with an organophosphorous insecticide used in the normal regime for control of mosquito larvae (Ward et al., 1976). However, uncaged plots treated identically showed significant reductions, compared to control plots, in population density of fiddler crabs over the summer. These results indicated that the insecticide itself only affected the crabs sublethally, since the crabs suffered unusual mortality only when they were subjected to both insecticide and predation. Laboratory studies (Ward and Busch, 1976) provided evidence that the sublethal effect of the insecticide was a slowing or disappearance of the fiddler crabs' normal escape behavior.

It should be evident from the factors and examples considered above that simple dose–response laboratory or short–term field experiments cannot form a basis for predicting field effects of stress factors on communities, populations, or even individuals. The interactions among the effects of an anthropogenic stress and field-specific phenomena, such as physicochemical variability, accumulation effects, behavior of species, susceptibility of different life stages, physiological condition, and a host of intra- and interspecific interactions, can be studied by simultaneous and interacting use of explicit modeling, field experiments, and laboratory experiments of appropriate complexity. Modeling has an important role in selecting aspects for laboratory and field studies that are probably most relevant in determining the outcome of the model's predictions. The complexity of laboratory and field experiments is important, since often inclusion of additional compartments (e.g., sediment in an algal culture system) or factors (e.g., physiochemical variables, different developmental stages of an organism) allow us to describe some important functional relationships that may be operating under natural conditions. As laboratory experiments increase in system complexity by the addition of compartments, physical models of ecosystems (or microcosms) are produced.

III. PHYSICAL MODEL SYSTEMS

In one sense, every laboratory experiment is a physical model system, but we tend to think of physical model systems as fairly complex and containing a number of the components present in the real field system being mimicked. There is, in fact, a variety of possible testing systems along a continuum of increasing complexity and naturalness (and usually size) as follows: (1) unispecific laboratory systems comprising a test organism and only those conditions necessary to keep the organism alive; (2) unispecific systems with one or two added components, intended specifically to examine the effects of those components on the test organisms; (3) laboratory or outdoor microcosms, that is, complete closed artificial ecosystems, that are small-scale models of natural ecosystems; (4) laboratory or field testing systems that are enclosed parts of larger natural ecosystems; (5) open natural ecosystems. Generally, the system types 2–4 are thought of as physical model systems or microcosms. Although the term microcosm is used by some to refer only to system type 3, we will assume the more general meaning.

The usual applications of physical model systems are to study autecological questions in a more realistic system than in unispecific laboratory cultures and to study synecological characteristics, such as interspecies interactions, community phenomena, and ecosystem function, under more controlled conditions than in a field study. Physical model systems allow holistic studies and have the additional advantages of being practical, controlled, replicable, and manipulable. They are usually practical because study and sampling of a series of laboratory vessels or enclosures is much easier than field sampling, particularly for aquatic ecosystems. Many environmental parameters, in particular climatological conditions, can be controlled (held constant, varied systematically, or varied at random). In some cases, sampling bias due to consuming the species sampled, as described by Falk (1974) for field situations, can be avoided in physical model systems, since replicate microcosms can be terminated at the time of sampling. A number of replicates can be set up, in contrast to most field situations, so that effects of manipulations can be examined statistically with greater ease and studied with more versatile experimental designs. It should be mentioned here, however, that as physical model systems become larger and more complex and realistic, the number of replicates practicable decreases drastically. A case in point is the CEPEX (controlled ecosystem pollution experiment) program. Very large (approximately 1700 m³ volume at full scale), complete ecosystem enclosures are used to study a variety of phenomena in a marine environment. For obvious logistic reasons, replication is limited to four systems (Takahashi et al., 1975; Koeller and Parsons, 1977). As points of comparison, one study mimicking marine communities with 150-liter laboratory microcosms consists of twelve replicates (Oviatt et al., 1977b), and many studies using simpler microcosms in standard laboratory flasks of a few hundred milliliters include as many as hundreds of replicates (Cooke, 1971). In any case, microcosm systems usually offer more opportunity for replication than natural field situations. Experimental manipulations are usually greatly facilitated by use of microcosms as opposed to field experiments, since microcosms are often small, generally not exposed to weather, in close proximity to each other, and usually not subject to the dilution effects that may follow some experimental manipulations in the field. Further, dangerous or potentially destructive experimental treatments may be performed on microcosms without risking damage to natural systems.

Obviously a number of problems, as well as advantages, are associated with the use of microcosms as hypothesis-testing systems. A central problem is that of level of mimicry demanded. How similar should a microcosm be to the real system? Further, how can we judge whether microcosm responses are sufficiently close to natural system responses? Oviatt et al. (1977b) argue convincingly that microcosms need not be identical to the system being modeled, but that microcosms should be generally similar to the real system in trophic structure, general taxonomic composition, productivity, rates of material cycling, and responses to perturbations. They add that microcosm communities need to be self-maintaining over a time span appropriate to the phenomena being studied. In terms of evaluating the responses of microcosms, Oviatt *et al.* suggest and use multivariate statistical methods, including canonical analysis and correspondence analysis, as a means of obtaining an objective statistical description of overall system state from data on a large number of different ecosystem parameters. It should be noted that these considerations apply to use of microcosms for holistic studies rather than when microcosms are used in a more reductionistic manner, including only a few selected components.

A number of investigators have constructed microcosms that respond similarly or identically to replicate microcosms and to the natural system being modeled according to their own evaluative criteria. Oviatt *et al.* (1977b) describe the fidelity and replicability of their microcosms of a marine bay ecosystem over a 6-month study period. Takahashi *et al.* (1975) show that four large natural marine water columns enclosed by plastic containers behaved biologically very similarly to each other and to the water outside the containers during a 30-day study period. Fidelity and replicability of microcosms over a number of seasons was reported for six laboratory model streams using natural stream components (McIntire and Phinney, 1965; Davis and Warren, 1965; McIntire *et al.*, 1964; McIntire, 1969).

It should be noted that the problems of evaluation of performance are certainly not restricted to microcosms as hypothesis-testing systems. Mathematical models and field experimentation have similar evaluative problem areas, as has been mentioned previously. Often, each technique has a number of benefits to be considered irrespective of exact mimicry or predictive power. Similarly to mathematical modeling, physical modeling involves explicit precise formulation of many assumptions and gives rise to modified or new hypotheses during the modeling process (design of the microcosms).

In addition to performance evaluation, there are a number of other problem areas in microcosm studies. Attention should be devoted to scaling the systems appropriately both physically and biologically (Perez *et al.*, 1977; Oviatt *et al.*, 1977b). In some cases, the wall effects of small enclosed systems can alter system responses. For example, added polutants often adsorb onto container walls, which typically represent most of the surface area available in an enclosed system. Further, microcosms lack the spatial heterogeneity often present in natural systems. Takahashi *et al.* (1975) argue that although this can be a problem, in their pelagic marine ecosystem, spatial heterogeneity appears to be due mainly to physical parameter variability rather than to differences in biological processes. Thus, useful information can be obtained by study of changes with time rather than only by study of changes with time over space.

In many microcosm studies, the higher trophic levels, or at least the larger animals, are excluded for obvious reasons. This omission may produce unknown response artifacts in some cases, particularly for ecosystems with dominant species of large body size. Very large physical models, such as the CEPEX systems, can accommodate some of the larger organisms and several trophic levels (Koeller and Parsons, 1977).

Other aspects of microcosm ecology, including metabolism, succession, and regulation of aquatic laboratory microcosms, are reviewed by Cooke (1971).

For most environmental impact studies, a large-scale long-term physical modeling effort of the CEPEX type is impossible on practical grounds. However, small-scale microcosm studies should be seriously considered as one of a number of hypothesis-testing methods to approach either autecological impact questions (e.g., Gillott et al., 1975) or explore holistic impacts of pollution or other factors. An interesting illustration of the latter use is the study by Oviatt et al. (1977b) of the effects of sewage addition to aquatic microcosms as a model of sewage addition to Narragansett Bay. Twelve replicate microcosms, each containing 150 liters of whole water samples and natural sediment communities from the bay, were constructed from 12 plastic tanks in a flow-through water system. The microcosms were carefully scaled to the bay physically and biologically. The 12 tanks were studied for 15 days to check on replication characteristics. Then 9 tanks were treated with three levels of treated urban sewage during a 3-month period, while the remaining 3 tanks were maintained as controls. The microcosms responded to the gradient of sewage treatments by developing characteristics similar to those found along a gradient from the Providence River, the point of entry of over 90% of all the sewage input to Narragansett Bay, to the mouth of the bay. Analysis of these characteristics included time series data and multivariate techniques (these methods can incorporate a large number of parameters simultaneously in a statistical description of the state of the system). After the 3-month treatment period, sewage input was discontinued and the recovery response of the microcosms was monitored for 2 months. During this time, all the microcosms became increasingly similar. The tanks that had been subjected to higher levels of sewage, however, were still far from complete recovery by the end of 2 months. These results in a 6-month experimental period are good indications that carefully designed and analyzed microcosm studies can be profitably used for environmental impact prediction experiments.

IV. STATISTICAL DESIGN OF LABORATORY EXPERIMENTS

It is important to formulate a statistical design for analysis of data before any experimental (laboratory or field) data are collected. The classic story of the biologist presenting a statistician with a mass of experimental data and having the statistician throw up his hands in despair will hopefully soon become a thing of the past. Unfortunately, the inclusion of statistics courses in biology curricula does not seem to have solved the problem of lack of appropriate design in many biological experiments. This is perhaps because such courses tend to devote most of their time and emphasis to the mechanics of statistical estimates and tests rather than to their applicability characteristics. In what follows, I will attempt to present some guidelines to facilitate choice of a statistical design for the types of biological experiments that occur commonly in environmental impact studies. I will assume an elementary knowledge of statistics such as that gained from a university statistics course for science students. It should be noted that this discussion applies equally well to laboratory or field experimental design; however, inappropriate design in a laboratory experiment is even less excusable than in a field situation, since laboratory experiments usually are more manipulable and easily replicable than field experiments.

There are two reasons why we may perform an experiment. First, we may wish to estimate a parameter or a relationship, such as the mean or standard deviation of a variable, or the regression coefficient of variable Y on variable X. Second, we may want to test an hypothesis (statistical inference) such as "the frequency of attribute A in the observed sample does not differ significantly from the expected frequency of attribute A according to a certain model," "the average value of X treated with compound A does not differ from the average value of X treated with compound B," or "the effect of factor A on X does not differ from the effect of factors A and B combined on X." Estimating parameters and relationships is fairly straightforward; many textbooks are available for reference (e.g., Sokal and Rohl, 1969; Schor, 1968; Woolf, 1968). These texts can also be consulted for details of the hypothesis-testing methods discussed below.

A large number of statistical tests and analytical methods are available for use in experimental data analysis. Table 5.1, an expanded and amended version of a table presented by Siegel (1956), displays the applicability and other characteristics of some of the most commonly used estimating statistics and statistical tests. Table 5.1 should be referred to frequently to complement the discussion that follows. One of the first criteria to examine when choosing a statistical test (and thus designing an experiment) is the type of scale of measurement that the data will represent (Stevens, 1946). If the data will be values on nominal (classificatory) or ordinal (ranking) scales, then the operations of arithmetic should not be performed on the data; thus, means and standard deviations should not be computed for such data. Other measures of central tendency and spread, appropriate for these measurement scales, are shown in Table 5.1. Correspondingly, nonparametric statistical tests should be used to analyze data from measurements on nominal and ordinal scales; such tests do not require use of means and standard deviations.

Often, biological data are measurements on interval or ratio scales (see Table 5.1). In such cases, parametric statistical tests should be used to take full advantage of all the information contained in the data, unless there is some reason to believe that the population from which the data will be drawn has characteristics that violate the major assumptions of the parametric test involved. Most often, these assumptions are normality or near-normality of distribution of variable values, which is almost never checked for, and homogeneity of variance, which is sometimes checked for by doing an F_{max} test or Bartlett's test on pilot data (Sokal and Rohlf, 1969). These assumptions about parameters of the distribution underlying the statistical tests, and thus about the distribution of the data that the test distribution is modeling, are responsible for the term "parametric" tests. As long as sample sizes are fairly large, the common parametric tests are robust to violations of their assumptions. However, at small sample sizes, this is not the case, and nonparametric tests should be considered as alternate possibilities even for measurements on interval and

TABLE 5.1.

Characteristics of Some Commonly Used Estimating Statistics and Statistical Tests[a]

Scale	Defining relations	Examples	Some appropriate statistics	Some appropriate statistical tests and analytical methods	
Nominal (classificatory)	1. Equivalence	Species A, B, or C Strain "green eyes," or "long winged" Habitat "forest," "grasslands," or "ecotone"	Mode Frequency Contingency coefficient	Binomial test χ^2 Sign test	Non-parametric
Ordinal (ranking)	1. Equivalence 2. Greater than	Size class "1," "2," "3," or "4" Age class "adult," "juvenile," or "1-year old" Response "strong," "medium," or "weak"	Median percentile Spearman r_s Kendall τ Kendall W	Binomial test χ^2 Sign test Runs test Mann–Whitney U test Friedman or Kruskal-Wallis ANOVA	
Interval (measurement)	1. Equivalence 2. Greater than 3. Known ratio of any two intervals	Temperature scales pH	Mean (arithmetic) Standard deviation Pearson and multiple product-moment correlations Regression (probit, multiple, etc.) co-efficients	χ^2 Student's T test ANOVA Multivariate analysis Cluster analysis	Parametric

(continued)

TABLE 5.1. (Continued)

Scale	Defining relations	Examples	Some appropriate statistics	Some appropriate statistical tests and analytical methods
Ratio (measurement with a true zero)	1. Equivalence 2. Greater than 3. Known ratio of any two intervals 4. Known ratio of any two scale values	Weight Population density Number of organisms Salinity % Cover	Mean (arithmetic) Standard deviation Pearson and multiple product-moment correlations Regression (probit, multiple, etc.) coefficients Geometric mean Coefficient of variation	χ^2 Student's T test ANOVA Multivariate analysis Cluster analysis

[a] Modified and expanded from Siegel (1956).

ratio scales. Siegel (1956) conservatively recommends $n = 6$ as the sample size at which and below which only nonparametric tests should be used, unless we know from previous data on the same or similar populations that we are dealing with normal distributions and homogeneous variances. This sample size is similar to that recommended in Chapter 4 as a minimum for field experiments ($n = 5$). However, in laboratory experiments, it is usually possible to plan considerably larger sample sizes and to use parametric statistics with fair confidence that their assumptions are not being grossly violated.

Another situation that would lead to choosing a parametric test is when data from complex designs (many factors included) need to be analyzed, since there are no well known nonparametric substitutes for multifactor ANOVA. This would be the case for some of the more complex laboratory studies that are useful in environmental impact work (see Section II).

Finally, another criterion for choosing between parametric and nonparametric tests is how small an effect or a difference is to be detected. When small differences need to be detected, generally parametric tests are more likely to detect them (they usually have greater power). An alternative is to increase the sample size and use a less powerful nonparametric test. Siegel (1956) discusses this topic in terms of the power and the "power-efficiency" (power gained by increasing sample sizes) of a statistical test.

Since, in many cases, biological data from laboratory experiments are measurements on ratio scales, include considerable replication, and can represent complex designs, parametric statistical tests are used in their analysis. The choice of the specific test to be used depends largely on the purpose of the analysis (e.g., estimate location, dispersion, establish functional relationship, estimate association, etc.), the measurement level attained (type of data), and the number of variables and number of samples (conditions) involved. Some of the most commonly used parametric tests are listed in the last column of Table 5.1. Sokal and Rohlf (1969) provide a useful Tabular Guide to Statistical Methods in an appendix that helps in selecting an appropriate statistical method. Another work that is of help for this purpose is that of Linton and Gallo (1975), which presents branching diagrams for method selection; it also has capsule descriptions and alternate appropriate tests for each type of analysis, as well as slightly lengthier material on set-up, computations, and interpretation for each method. Curvilinear regression, a method not covered by either of the two references just given, is discussed by Poole (1974).

A relatively new and more complex group of statistical methods is available that can incorporate a number of variables simultaneously such that a system state can be described in laboratory or field experiments. These are methods of multivariate analysis, and generally use a matrix of correlation coefficients among the many measured variables as a starting point for the analyses. Cluster analysis and other similarity methods *group the variables* according to their correlation coefficients and are used commonly in describing complex communities of organisms (e.g., Foreman, 1977). Another group of multivariate methods, called factor analytic methods, *identifies a few underlying factors* that explain a large part of the variance and covariance of the original variables (e.g., Oviatt *et al.*, 1977b; Miracle, 1974). Pielou (1969) discusses the basic features of these methods. An especially interesting example of the use of multivariate methods in environmental impact studies is the study by Oviatt *et al.* (1977a) of variation and evaluation of coastal salt marshes. This study compared ten salt marshes differing markedly in degree of human influence. A large number of biological variables were measured in at least five replicates in each marsh, an effort that occupied a large number of people over a 2-year period. The data indicated that differences found among the marshes were small compared to the differences from place to place or from time to time within each marsh; further, there was little intercorrelation among the parameters, such that, for example, a marsh with a high standing crop of grass did not necessarily have more abundant animal populations than one with a low standing crop. Since individual parameters and univariate statistics did not allow separation of the marshes into categories of any sort, a multivariate statistical method (correspondence analysis) was used to examine the whole data set to compare the marshes and identify the factors that might allow their grouping. The analysis showed that seven out of the ten marshes were quite similar to each other (see Fig. 5.1). The three remaining marshes were not similar to each other. Further, the variables differentiating these three marshes from the larger group were mainly measures of fish populations, that are notoriously patchy. Thus, these variables were judged to be of little value in separating the marshes. Since the analysis did not allow construction of groups of marshes and isolation of factors that allowed

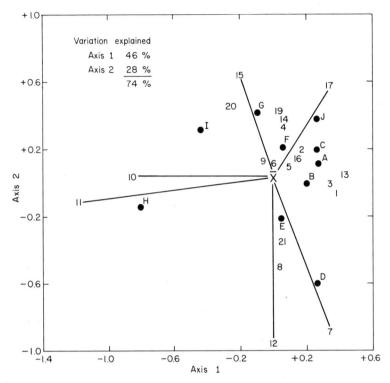

Fig. 5.1. Similarity of ten marshes as described by correspondence analysis. Both the location of the individual marshes (capital letters) and each variable (numbers) are plotted. Similar marshes are adjacent; dissimilar ones are distant. Variables that characterize a particular marsh are proximal to the marsh location. Variables found at extreme distances from the mean variable location ($\bar{X}$) should be good indicator variables for separating marshes. In this case most such variables are fish population measures, which are notoriously patchy (Nos. 7, 10, 11, 12, 15). (From Oviatt et al., 1977a.)

their grouping and separation, even with this considerable sampling effort, the study concludes that ecological rating or classification systems will not be a useful tool to manage coastal habitats. This use of multivariate techniques makes an interesting contrast to their use in evaluating the response of microcosms to disturbance (Oviatt et al., 1977b) (see Section III). In the latter case, multivariate analysis clearly separated different sewage treatment levels, probably due to the replicability (similarity) from microcosm to microcosm in the untreated condition.

V. MEASUREMENT OF TOXICITY

Environmental impact studies often deal with estimating the effects of addition of a toxic substance to a community of organisms. In the past a measure called the LD_{50} of various species has been used as an indicator of toxicity. The LD_{50} (dose that kills 50% of the animals, or dose lethal to 50%) or LC_{50} (concentration lethal to 50%) is obtained by exposing test organisms to a range of concentrations of the toxicant in the laboratory and plotting the percent mortality versus the concentrations (Fig. 5.2). The concentrations are plotted as logs of the concentrations and the percent mortalities are plotted as probits. The probit transformation consists of finding the value of the normal deviate (value of the abscissa in a normal curve graph with variance equal to 1) for a given proportion (corresponding to the percent mortality) of the area under a normal curve and adding 5.00 to that normal deviate value (see Fig. 5.3). The normal deviate is used be-

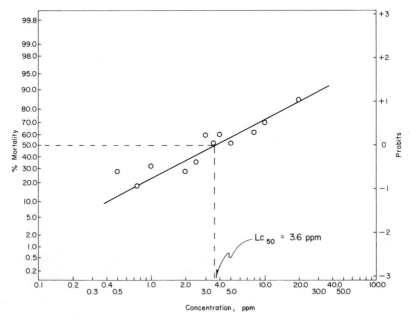

Fig. 5.2. Probit analysis graph showing the concentration–mortality response relationship. Although commonly only LC_{50}'s from these analyses are reported, the slopes of the graphs are of critical importance (see Fig. 5.5).

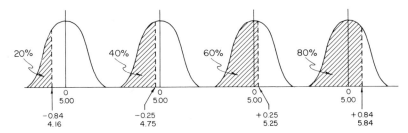

Fig. 5.3. Probit transformation. For a given percentage of the area under the normal curve (e.g., 20%) the corresponding normal deviate is given (−0.84) and 5.00 is added to the normal deviate (result equals 4.16). Hence the probit is a normal variate with mean 5.00 and variance 1. (From Li, 1964.)

cause it is assumed that the tolerance of the test animals to the toxicant is normally distributed. The constant 5.00 is added to avoid working with the negative numbers that would result for mortalities lower than 50% if the normal curve mean were 0. As a result of these transformations, a linear graph is obtained so that linear regression can be used to construct an appropriate line. Further details of the procedure can be obtained in Li (1964) and particularly in Finney (1964).

The LC_{50} data discussed above are obtained by exposing test populations of animals to the toxicant concentrations for a specified period of time (48- or 96-hour LC_{50} data are common). Obviously, the response will be different for different times of exposure, and the rate of change of the response varies for different species and different physiological and testing conditions of a species. For each experimental situation, a pilot test should be run to determine an appropriate exposure period. Standard exposure periods should not be adopted without specific information on the species in question and the physiological and test conditions, since in one situation most of the mortality may result in the first 24 hours, while in a different situation it may result after 48 hours or even longer. Brown (1973) pursues this topic in some detail.

The construction of graphs similar to Fig. 5.2 for a number of exposure times would lead to a family of concentration–mortality curves. Similarly, a number of time–mortality curves can result from experiments in which the time required to produce various percentage mortalities at specified toxicant concentrations is measured (see Fig. 5.4). The latter approach is probably not as useful as determination of concentration–mortality curves, especially when the appro-

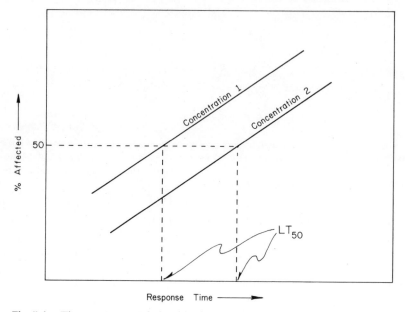

Fig. 5.4. Time–response relationship for two different concentrations of a toxic material. The time required to produce various percentage mortalities at specified concentrations is measured.

priate exposure period is chosen by pilot experiments. The concentration–mortality graph allows objective statistical definitions of effective concentrations and their associated errors, rather than the subjective estimates of effective concentrations or dosages obtained from graphs of mortality versus response times for specified toxicant concentrations.

Although testing the toxicity of substances to organisms can be profitably used in environmental impact studies, many past toxicity studies have been at best irrelevant to environmental impact prediction. First, the use of isolated LD_{50} data for comparisons across species or across toxicants is dangerous, since the slopes of the graphs of mortality versus dosages may differ greatly from species to species for the same compound and for different toxicants for the same species. Thus species A may show a higher LD_{50} than species B, but a lower LD_{80} than species B (see Fig. 5.5). Second, the experimental conditions for toxicity testing are so different from any complex natural situation that substantial design efforts are needed to gain some ability to extrapolate from laboratory results. Design of toxicity testing experiments must take into account the effects of different life stages of a species, any sublethal effects of toxicants leading to population changes, and interactions of toxicant effects with biological and physicochemical factors. One approach that has been used to study some aspects of this complexity is to measure the effects of several simultaneously manipulated variables in the laboratory experiment. For example, Angelovic *et al.* (1969) constructed a three-dimensional plot of mortality as a function of ionizing radiation, salinity, and temperature for a killifish (*Fundulus heteroclitus*) (see Fig. 5.6). McLeese (1956) used a similar approach to define the mortality regions of the American lobster when exposed to simultaneously varing temperature, salinity, and oxygen (see Fig. 5.7). Another approach to increase

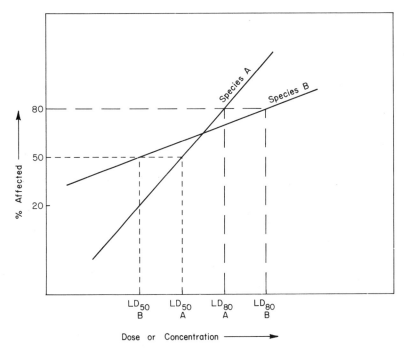

Fig. 5.5. Differing slopes of the dose–response curve may lead to markedly different relative effects on different species at different dose ranges. Thus comparisons of LD_{50} values to express relative sensitivity to a toxic material may be misleading. In addition, different species may have differentially sensitive immature stages and sublethal effects.

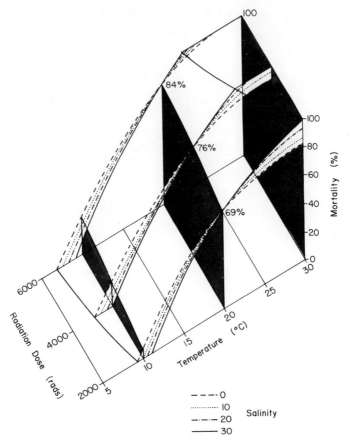

Fig. 5.6. Predicted effects of temperature, salinity, and radiation dose on the percentage mortality of mummichog (*Fundulus*) 20 days after irradiation. The different effects of the three radiation doses are most apparent at 20°C or where the mortality at any given salinity reaches 100%. Note the temperature–salinity interaction shown by increased mortality with a decrease in salinity at temperatures below 20°C and decreased mortality with a decrease in salinity at temperatures above 20°C. (After Angelovic *et al.*, 1969.)

the usefulness of laboratory toxicity testing is to investigate the relevant exposure conditions and life stages involved by methods such as field experimentation and mathematical modeling. In interaction with these methods, laboratory toxicity testing can be designed to relate closely to expected field conditions (see Section II).

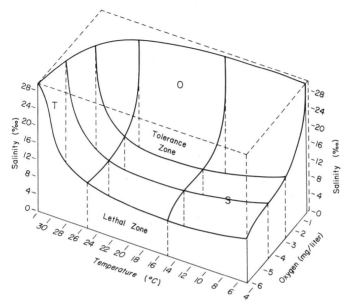

Fig. 5.7. Mortality of the American lobster as a function of temperature, salinity, and oxygen: the diagram shows the boundary of lethal conditions. T, region in which temperature alone acts as a lethal factor; S, region in which salinity alone acts as a lethal factor; O, region in which oxygen alone acts as a lethal factor. The smooth curves form the boundary between the lethal zone below and the tolerance zone above. Thus the diagram can be used to assess the lethal effects of temperature, salinity, or oxygen with respect to the other two factors. For further interpretation see McLeese (1956). (From McLeese, 1956.)

REFERENCES

Angelovic, J. W., White, J. C., Jr., and Davis, E. M. (1969). *Symp. Radioecol. USAEC Conf., Oak Ridge, Tennessee* pp. 420–430.

Brown, V. M. (1973). *In* "Bioassay Techniques and Environmental Chemistry" (G. E. Glass, ed.), pp. 73–96. Ann Arbor Sci. Publ., Ann Arbor, Michigan.

Cooke, G. D. (1971). *In* "The Structure and Function of Fresh-Water Microbial Communities" (J. Cairns, Jr., ed.), Res. Div. Monogr., No. 3, pp. 47–85. Virginia Polytech. Inst. State Univ., Blacksburg.

Davis, G. E., and Warren, C. E. (1965). *J. Wildl. Manage.* **24**, 846–871.

De Coursey, P. J., and Vernberg, W. B. (1972). *Oikos* **23**, 241–247.

D'Itri, F. M. (1973). *In* "Bioassay Techniques and Environmental Chemistry" (G. E. Glass, ed.), pp. 3–71. Ann Arbor Sci. Publ., Ann Arbor, Michigan.

Falk, J. H. (1974). *Oikos* **25**, 374–378.

Finney, D. J. (1964). "Probit Analysis," 2nd Ed. Cambridge Univ. Press, London and New York.

Foreman, R. E. (1977). *Helgol. Wiss. Meeresunters.* **30**, 468–484.

Gillott, M. A., Floyd, G. L., and Ward, D. V. (1975). *Environ. Entomol.* **4**, 621–624.

Koeller, P., and Parsons, T. R. (1977). *Bull. Mar. Sci.* **27**, 114–118.

Li, C. C. (1964). "Introduction to Experimental Statistics." McGraw-Hill, New York.

Linton, M., and Gallo, P. S., Jr. (1975). "The Practical Statistician: Simplified Handbook of Statistics." Brooks/Cole Monterey, California.

McIntire, C. D. (1969). *Proc. Eutroph. Biostimul. Assess. Workshop, Berkeley, Calif.* pp. 146–157.

McIntire, C. D., and Phinney, H. K. (1965). *Ecol. Monogr.* **35**, 237–258.

McIntire, C. D., Garrison, R. L., Phinney, H. K., and Warren, C. E. (1964). *Limnol. Oceangr.* **9**, 92–102.

McLeese, D. W. (1956). *J. Fish. Res. Board Can.* **13**, 247–272.

Miracle, M. R. (1974). *Ecology* **55**, 1306–1316.

Oviatt, C. A., Nixon, S. W., and Garber, J. (1977a). *Environ. Manage.* **1**, 201–211.

Oviatt, C. A., Perez, K., and Nixon, S. W. (1977b). *Helgol. Wiss. Meeresunters.* **30**, 30–46.

Perez, K., Morrison, G., Lackie, N. F., Oviatt, C., Nixon, S., Buckley, B., and Heltshe, J. F. (1977). *Helgol. Wiss. Meeresunters.* **30**, 144–162.

Pielou, E. C. (1969). "An Introduction to Mathematical Ecology." Wiley, New York.

Pimentel, D. (1971). "Ecological Effects of Pesticides on Non-Target Species." Exec. Off. Pres., Off. Sci. Technol., Washington, D.C.

Poole, R. W. (1974). "An Introduction to Quantitative Ecology." McGraw-Hill, New York.

Ryther, J. H. (1954). *Biol. Bull. (Woods Hole, Mass.)* **106**, 198–209.

Schor, S. (1968). "Fundamentals of Biostatistics." Putnam, New York.

Siegel, S. (1956). "Non-Parametric Statistics for the Behavioral Sciences." McGraw-Hill, New York.

Sokal, R. R., and Rohlf, F. J. (1969). "Biometry." Freeman, San Francisco, California.

Stevens, S. S. (1946). *Science* **103**, 677–680.

Stickel, L. F. (1973). *In* "Environmental Pollution by Pesticides" (C. A. Edwards, ed.), pp. 254–312. Plenum, New York.

Takahashi, M., Thomas, W. H., Seibert, D. L. R., Beers, J., Koeller, P., and Parsons, T. R. (1975). *Arch. Hydrobiol.* **76**, 5–23.

Vernberg, W. B., and Vernberg, J. (1972). *U.S. Fish Wildl. Serv., Fish. Bull.* **70**, 415–420.

Ward, D. V., and Busch, D. A. (1976). *Oikos* **27**, 331–335.

Ward, D. V., Howes, B. L., and Ludwig, D. F. (1976). *Mar. Biol.* **35**, 119–126.

Woodwell, G. M. (1967). *Sci. Am.* **216**, 24–31.

Woolf, C. M. (1968). "Principles of Biometry." Van Nostrand, Princeton, New Jersey.

6

Some Examples

I. COMMON TYPES OF BIOLOGICAL ENVIRONMENTAL IMPACT STUDIES

A major motivation for writing this book was the fact that most existing biological impact studies do not use the approaches advocated here. There are two major classes into which the majority of previous impact studies fall: the busy taxonomist approach and the information broker approach.

In the busy taxonomist approach, an environmental impact study consists of a lengthy, costly survey of everything present in the system to be impacted. A considerable volume of taxonomic work is performed, often including identification of specimens by experts of every description. In some of these studies, there is an attempt to describe some functional characteristics, but always in the unimpacted state of the system. The basic premises underlying this approach are that (1) an extensive set of field-collected data is necessary to understand the system and that (2) knowing what is present in the system and perhaps something about how it operates in its current state will allow us to guess how it will operate when changed by development actions. Our discussions, throughout this book, of a variety of examples of unexpected ecosystem effects (biological magnification of DDT, effect of nutrients from duck manure on oyster survival in Long Island bays, mortality of fish species from sudden cessation of heated effluent in the winter, to name a few examples) clearly invalidate these premises. In all of these cases, an experimental approach of some sort (on part of the system, on microcosms, etc.) would have stood a much better chance of predicting the eventual effect, even if the data collected had been scantier and less comprehensive taxonomically than in most of the studies belonging to this class. Needless to say, this type of study is expensive, time-consuming, and not cost-effective in predicting possible impacts. The only features to recommend this type of study is that the investigators are usually sincere in their belief that this is the best way to study these matters rather than performing the studies this way as a means of rapid profit in business.

In the information broker approach, practiced by numerous enterprising consulting firms, quite the opposite is true with regard to motivation. A prime objective is to obtain as many contracts as possible and satisfy the terms of reference in the shortest possible time. The result is a literature and other-investigator survey of existing information on the ecological system, a presentation of alternative plans and schedules for the proposed developments, and a series of guesses as to how the ecological system might respond to each development alternative. If there is any original data gathering at all, it is of the descriptive survey kind rather than manipulative, and even that usually suffers from lack of adequate sampling replication and design as a result of the haste involved. In short, the information broker approach consists of describing the proposed development based on information supplied by the client and of doing a literature/field survey of the ecological system involved in its present state and guessing at possible impacts. Yet this kind of information brokerage, containing no new analyses or new experimentation, costs millions of dollars every year in North America; consulting firms of this type abound. The approach is, of course, again based on the premise that descriptive information in the present state allows prediction of system behavior in changed states. In a few cases, impact studies of this type may contribute by reporting on impacts discovered elsewhere in similar systems in response to similar developments.

II. SOME COMMENTS ON EFFECTIVE ENVIRONMENTAL IMPACT STUDIES

No one example presented in the sections that follow uses all the approaches advocated here, and each study has individual peculiarities due to the kind of development or impacting action involved, the kind of ecological system involved, the terms of reference for the study (effects on whole ecosystem, on target species, on economically important species, etc.), the amount of previous data available, etc. However, the three examples presented illustrate a substantial number of the aspects discussed and advocated here as effective techniques for biological impact assessment. All three studies produce results that can be used to help make current management decisions *and* to make critical priority lists of research needed to further reduce uncertainty. All deal with impact prediction by performing the development actions in one form or another [manipulate a small portion of the ecological system, compare an undeveloped system with a similar but developed (impacted) system, or manipulate a simulation model of the system] rather than by guessing at system responses on the basis of the unimpacted system only. All three are closely focused on collecting and using only data that are directly related to prediction of impact on the components of interest but at the same time deal primarily with system properties in the overall analysis. The range of time scales for the studies should be noted: the first study lasted 3 years, the second study lasted about 3 months, and the third study lasted about 3 days.

The first example is described in greater detail than the other two because my involvement in the project allows me to cover a lot of the rationale, the decision processes, and the many problems encountered as well as the work done and the results obtained.

III. IMPACT OF INSECTICIDES ON SALT MARSH ECOLOGY

A. Scope

This study was carried out by a varied natural sciences team at Rutgers University. It was funded by the Mosquito Control Commission of the State of New Jersey. The intended study period was 5 years; the actual study period was 3 years (June 1972–June 1975), as funding was no longer available after that time. The overall goal of the project was to determine whether the mosquito control chemicals used in New Jersey presented environmental risks. Elucidation of the nature and magnitude of these risks would provide information useful for decision making on insecticide use. A complete analysis of costs and benefits (biological, economic, and social) of insecticide use was not considered to be in the scope of this study by either the funding agency or the investigating team. The objective of this effort was to estimate only possible biological environmental costs of insecticide use. Since it was necessary to narrow the consideration to a single ecosystem, the salt marsh ecosystem was selected for study. This choice was made because salt marshes represent the majority of mosquito insecticide application sites in New Jersey, because they are thought to be valuable natural ecosystems, and because they are ubiquitous (200,000–250,000 acres are still present in New Jersey) in the east coast of North America. Similarly, a single insecticide (Abate or temephos) was chosen initially for study. Two different formulations of temephos and another, more toxic insecticide, Dursban (chlorpyrifos), were eventually evaluated in the study. Temephos was chosen because it is an organophosphorous compound that represents the class of insecticides most commonly used in North America following abandonment of the organochlorine compounds for most uses. Also, it is the compound most commonly used for mosquito control in New Jersey. Temephos is used as a larvicide in salt marshes. It is applied on marshes either as an emulsion in oil or as a granular formulation on clay granules. It was obvious from the outset that such restrictions on the scope of the study would seriously limit the applicability of the results to different situations. However, the bounds on the scope of the project were necessary to define a realistic experimental system.

B. Experimental Approach and Focusing Mechanisms

The approach adopted by the study team was that experimental treatments of the marsh system with insecticides were necessary to estimate possible effects of insecticide use and that small parts of the natural ecosystem would be used as experimental units. Considerable thought was given to using a comparative and/or a monitoring approach (i.e., to find marshes that had been treated in the past and compare them to pristine marshes, or to study marshes that would be treated by the Mosquito Control Commission as part of their normal control program during the study period and compare those to untreated marshes). However, these approaches were rejected because natural variability from marsh to marsh was estimated to be too great to allow comparison across different marshes. Later empirical work in Rhode Island marshes (Oviatt et al., 1977) tends to support this evaluation. Thus, comparison of treated and control conditions was made within the same marsh, using experimentally applied treatments in small portions of the marsh.

The means chosen for focusing on critical aspects for study was to identify characteristics that seemed to play a major role in the overall ecological function of the marsh being studied. Thus, dominant and key species and processes would be examined for their responses to insecticide treatments. Mathematical modeling was not used at this stage to focus on research priorities, since the intent at this point was to scan the marsh system for impacts of insecticide treatments so that a very large number of compartments and uncertainties would be involved in any attempt at modeling. Mathematical modeling would be of greater use applied to specific populations of interest following the general ecological study.

C. Planning Experimental Studies

The planning steps followed in designing the research program were as follows: (1) a rapid literature and field survey was made to describe the ecological characteristics of the marsh; (2) key characteristics and relationships were identified from the description of ecological structure and function and from other "importance" viewpoints (see Chapter 2, Section V); and (3) experiments were designed.

The initial literature and field study occupied only a month or two for most of the team members, and this period involved other activities (hiring personnel, buying equipment, etc.) in addition to the survey study. The survey was made only to provide a brief qualitative description of the ecosystem. The emergent marsh was dominated by the grasses *Spartina alterniflora* and *Spartina patens*. Other prominent emergent marsh organisms were insects and spiders; a number of bird species; benthic invertebrates including the fiddler crab *Uca pugnax*, the snail *Melampus bidentatus*, amphipods and isopods; a number of algal species; and bacterial/fungal communities. Additional major components found in marsh ponds and channels were the killifish or mummichog, *Fundulus heteroclitus*; the sheepshead minnow, *Cyprinodon variegatus*; the shrimp *Palaemonetes sp.*; and the widgeon grass, *Ruppia maritima*. A number of other components were present in the marsh system but were judged not to be as important (using the various kinds of importance) as the components listed above. The grasses were considered vital, since they produced the majority of the energy in the system and also were physical dominants (roots and blades provided substrate for most of the marsh ·biota). Because marshes are known to be primarily detritus-based ecosystems, detritivores, such as fiddler crabs, and bacterial/fungal communities were assigned considerable importance. Fiddler crabs were also considered potential "indicator" organisms in that they had previously been shown to be markedly sensitive to a number of pollutants (Ward *et al.*, 1976; Ward and Busch, 1976). The small fish species present were thought to be important both because they appeared to be dominant numerically and by biomass and because they are known predators of mosquito larvae, representing a natural control agent for the target species of the treatment being studied. Nontarget insects were also considered important for a number of reasons. Because of their relative physiological similarity to mosquitoes, susceptibility to mosquito insecticides might be suspected (particularly in aquatic insects, since the insecticides being tested were primarily larvicides acting in marsh ponds). Second, some nontarget insects are predators of mosquitoes, thus representing a natural control agent that may be negated by insecticide treatments. Last, insects represent most of the species diversity present on marshes. Bird species were ascribed human importance values (hunting use, aesthetic value) in addition to their roles in marsh ecology. Study of these biological components was to be accompanied by a variety of chemical studies on insecticide (and insecticide breakdown product) cycling and persistence in the ecosystem.

In addition to the primary effort involving field experiments, a variety of laboratory experiments was planned to examine hypotheses interactively with the field studies.

The design of the field experiments began by delimiting sampling areas in an unpolluted marsh site. The control and experimental plots were made large enough (0.5 hectare) so that, for most components, the total area expected to be sampled by the end of the experiments would not exceed 10% of the total area available. The plots were separated enough to prevent contamination of control plots from treated plots; these distances were checked by sampling for the insecticides in control plot water, sediments, and biota in both an initial trial application and during the course of the experiment. The spacing needed to prevent contamination of control plots was not too great to allow close matching of plots within the same marsh. The sampling areas were adequate for sampling most compartments of interest except for the birds, insects, and fiddler crabs. Birds are highly mobile, so that small plots are not sufficient to either reliably expose them, directly or indirectly, to the insecticide nor to count them within the area, since they often moved out of the plots when people were present. Therefore, a separate sampling area was selected for bird work, to minimize disturbance in the area, and there only sparrows (*Ammospiza maritima* and *A. caudacuta*) were censused. The sparrows tended to remain in the area until they were actively flushed out of the marsh grass, so that the presence of the person(s) sampling was not thought to affect the sparrow densities, particularly in comparisons of treated versus control plots. However, the sparrows did fly in response to flushing of the area by running a rope over the plot surface; thus they could be counted systematically as they flew up and away. The main hypothesis being tested in this experiment was whether a reduction in the utilization by sparrows of treated plots versus control plots occurred over the course of a summer. Such a short-term reduction in utilization of an area might be expected if the density of food items (insects, spiders, small crustacea) decreased as a result of insecticide treatments. An alternate hypothesis was that the food organisms may not immediately decrease in density, but become easier to capture as a result of insecticide-induced behavioral changes. Such a change may lead to increased utilization of treated plots, at least temporarily, before predation pressure began to reduce prey densities. Since data on sparrow densities and data on small arthropod densities could be collected, these two hypotheses would be tested. The results showed no significant changes in sparrow densities nor in small arthropod densities for any of the insecticides or formulations tested over one summer. Given available time and resources, it was not considered profitable to pursue larger spatial and time scale studies of effects on sparrow populations or any studies on other bird species, since their mobility precluded working in small sampling areas.

Insect sampling involved some problems in common with bird sampling, since many adult insects are highly mobile. No serious insect sampling was done in the first year of the project, since the plots were only 0.5 hectare each. However, methodology was developed during this period and 3-hectare insect sampling plots were established the following year. Three types of sampling were used in these plots: sweep-netting for grass-inhabiting insects, sampling with an open-ended cylinder in marsh ponds for aquatic insects, and UV-trap sampling in the center of the plots for nocturnal flying insects. Even in the 3-hectare plots, sampling the grass-inhabiting and the nocturnal flying insects was affected by movements of insects into and out of the plots. Results had to be interpreted with this consideration in mind. For example, more emphasis was placed on species thought to be resident marsh species, most likely to be affected by the insecticide treatments, than on species thought to be immigrants spending brief periods on the marsh. Even within resident populations 3 hectares may not have been enough to prevent sampling of "untreated" insects moving into the plots from adjacent areas shortly before sampling. Thus, aquatic insect sampling in marsh ponds was evaluated as a more reliable indicator of effects of insecticide treatment than grass-inhabiting or nocturnal flying insect sampling.

The original 0.5-hectare plots were also inadequate for study of fiddler crabs. The initial plots did not contain tidal creeks, since the terrain usually sprayed for mosquito larvae is relatively high marsh, containing ponds or pannes but not channels open to tidal flushing. However, fiddler crabs inhabit the banks of tidal creeks and are probably affected by insecticides to some extent, due to the relatively gross application localization possible by helicopter and/or to transport from treated areas by tidal water movement. Therefore, separate plots were established along selected creek banks for fiddler crab experiments.

Having selected the compartments for study and delimited tentative sampling areas, attention was next focused on sampling strategies. There was great variation in strategies adopted depending on the organisms being sampled, apparatus involved, cost of sampling, etc. The experimental design used for sampling marsh grasses will be described here as one example of designs used in the project; other examples can be found in Ward *et al.* (1976) and in Campbell and Denno (1976). The hypothesis being tested was whether treatments with temephos (and later also chlorpyrifos) changed the growth, as measured by standing crop at various times during the season, of *Spartina alterniflora* and *S. patens* from that observed in control plots in the same marsh. Since the stands of the two grasses within the plots appeared homogeneous within each grass type, random sampling was selected for use. The sample size ($n = 5$) was selected by examining pilot data (as described in Chapter 4, Section IV,D and Figs. 4.2 and 4.3), for the gain in stability of the estimates of mean and error as sample size increased from 1 to 10. Since we wished to compare samples from the treated plots to samples from the control plots (separately for the two species of grass) to test whether they came from the same population, and the population variance was not known, the t test for equal size of samples was used as a statistical test (in general, the homogeneity of variance assumed by the t test was met). It should be noted that these data can be analyzed in a number of other ways. One alternative is to test for seasonal (time) effects on the standing crop of the plants at the same time as for effects of insecticide treatments and any interaction between the two factors. This analysis can be done by use of ANOVA (as in Ward *et al.*, 1976; Campbell and Denno, 1976).

A variety of laboratory experiments was designed as a means of testing hypotheses that had been generated by the field/literature survey and by field experiments. This method of specifying questions for laboratory study minimized indiscriminate proliferation of laboratory experiments. For example, in designing laboratory studies on effects of temephos on fiddler crabs, a biologist might be tempted to measure mortality, changes in respiratory rate or in osmoregulatory function, effects on a variety of reproductive processes, effects at different life stages and under a variety of environmental conditions, effects on a large number of observable behaviors, etc. All of these are inherently interesting questions, but represent enough work for many times the resources and time available for the entire study. Thus, it is imperative to adopt some mechanism of selecting key questions for laboratory study. In this case, field experiment results were used to select these areas. A pilot study of field effects of temephos on fiddler crab populations (Ward and Howes, 1974) suggested that the observed reduction (about 20%) in fiddler crab density in treated plots when compared to control plots was not an immediate response to temephos application, but was only observed after the second application was made and/or 2 weeks had elapsed. The delay suggested two possibilities: (1) accumulation of temephos (by continued slow release from the granules on the marsh and/or by an additional application) was needed to reach effective lethal levels or (2) temephos had a primarily sublethal effect on the crabs, altering behaviors contributing to survival, which in time decreased the fiddler crab density. Laboratory experiments were designed to test those hypotheses: (1) the temephos content of crab tissues collected over the course of the field experiment was measured by gas chromatography and (2) the effects of temephos treatment on the escape reaction of fiddler crabs were studied in a controlled laboratory situation. The gas chromatographic study of the crabs collected after each quadrat was counted in the field showed that there was no buildup of temephos concentrations with time. The escape response, on the other hand, was impaired by exposure to temephos in laboratory tests (Ward and Busch, 1976). As final support for this mechanism of *Uca* population reduction, a field experiment was done in which changes in fiddler crab densities were followed in temephos-treated and untreated marsh plots and in treated and untreated plots that were caged over to reduce predation by marsh birds. Temephos significantly reduced the population density of fiddler crabs in the open test plots but not in the caged plots (see Fig. 6.1). Thus, both laboratory and field results indicated that temephos had a primarily sublethal effect on the crabs, the effect becoming lethal only after interaction with predation by birds on the marsh (Ward *et al.*, 1976).

D. Results

At the end of the 3-year study period, the major ecological effects of temephos (and some effects of chlorpyrifos) had been studied, some in greater depth than others. There was no discriminable effect during this time of the temephos or chlorpyrifos treatments on production by the *Spartinas*. No accumulation of the compounds in the grasses was found. *Spartina* seed germination as measured in the laboratory was unaffected by temephos treatments. The edaphic chlorophyll (chlorophyll *a* corrected for phaeopigments) concentration on the marsh was not affected by temephos treatments, at least for a period of 1 year after cessation of insecticide treatments. The effects of the insecticides on diatom community structure were studied, but no clear trends emerged; although diversity and evenness of diatoms decreased with temephos treatments in the initial marsh plots, the results did not replicate in the second set of marsh plots, nor did chlorpyrifos cause significant diatom diversity changes. Laboratory studies on a number of macroscopic and microscopic marsh algae in culture failed to show any significant effects of temephos treatments on their growth or survival. However, there was an interesting stimulatory effect of temephos on bacterial productivity in the laboratory algal cultures, which did secondarily reduce algal growth in a few species, whenever the bacterial growth was not controlled by neomycin additions. When the antibiotic was used to control bacterial

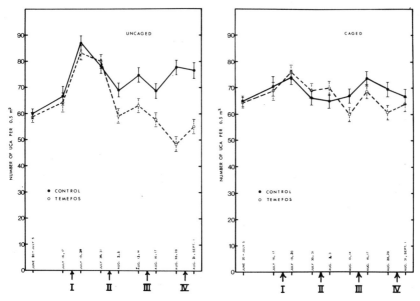

Fig. 6.1.   Effect of temephos treatments on population density of *Uca pugnax* in open marsh plots and in plots caged to reduce predation. The crabs' escape reaction was impaired by temephos treatments; in the open plots, this led to increased predation success and lowered *Uca* densities. In the caged plots, however, the effect of temephos remained sublethal, since the crabs were protected from predators. (From Ward *et al.*, 1976.)

growth, the algae showed no significant growth changes when treated with temephos in a wide range of concentrations.

Two attempts were made to study effects of temephos on decomposition processes in the marsh. Litter bags were used to measure possible changes induced by temephos in decomposition rates of the *Spartina* and laboratory methods were developed to measure the effects of temephos and other insecticides on the bacterial activity of marsh mud by monitoring aerobic bacteria respiration and anaerobic bacteria sulfate reduction. Problems encountered in all of these cases (method failures in the litter bag studies, insufficient investment of resources in the laboratory studies) precluded even tentative conclusions on effects of temephos on decomposition processes. This was seen as a major gap in the study, as decomposition processes are of great importance to detritus-based food webs; thus, this area should be of high priority for any future research. It was previously noted that temephos apparently stimulated bacterial growth in laboratory algal cultures; it is impossible to predict whether similar effects would be observed in the field on that basis.

The studies on the nontarget insects inhabiting grasses and flying nocturnally over the marsh did not show any significant effects of temephos or chlorpyrifos on several community structure measures. Both of these groupings were judged extremely variable and difficult to interpret when compared to the aquatic marsh insect community. Applications of both insecticides had no significant effects on the species richness, species diversity, and densities of the dominant, subdominant, and predatory aquatic insects in salt marsh potholes (Campbell and Denno, 1976). Although spiders were quite numerous in the marsh, their difficult taxonomy and the apparent extreme patchiness of their distribution discouraged investment of resources on spider studies.

As mentioned above, a number of benthic invertebrates inhabiting the emergent marsh were important from a number of points of view and were studied in some depth. A mixed community of macroepifaunal organisms including amphipods, isopods, the snail *Melampus bidentatus*, and a number of minor additional groups showed no significant effects of temephos or chlorpyrifos on densities or species composition. However, variability was considerable in these data. A more detailed study of *Melampus* populations showed no significant differences in population densities due to granular temephos or granular chlorpyrifos treatments, but a significant decline in population density with temephos in the emulsion formulation. The population density increased back to pretreatment levels when the insecticide treatments ceased. However, temephos residue content of the snails persisted for more than 5 weeks after the last treatment of the marsh with granules, while residues disappeared from snails exposed to emulsion treatments in less than 3 weeks after cessation of treatments (Fitzpatrick and Sutherland, 1976). The longevity of the residues in *Melampus* from granular temephos-treated plots may be due to slow release of the compound from the granules, so that the snails were exposed to a continuing supply of temephos even though the applications had ceased. Thus, these studies suggested that (1) some population decreases in *Melampus* were observed in the field, when emulsion formulations of temephos were used, perhaps due to interactions of sublethal toxic effects with ecological phenomena and (2) some persistence of temephos residues was observed in *Melampus* exposed to granular formulations in the field, perhaps due to slow release of the compound from granules. Both findings suggest that investigation of major predators of *Melampus* would be desirable to assess the potential for major changes in *Melampus* populations due to interactions between sublethal effects of temephos and predation and for movement of temephos residues through the marsh food web.

The fiddler crab *Uca pugnax* was another emergent marsh benthic invertebrate studied in some depth, since this species is thought to be important as a detritivore on the marsh, is present in large numbers, is an important structural dominant along creek banks due to its extensive deep burrowing activities, and appears to be extremely sensitive to pollution of any sort (Ward *et al.*, 1976). Field studies of population changes, under normal conditions and using caged plots to exclude most predators, and laboratory studies of mortality and sublethal effects of temephos supported the hypotheses that the compound impaired the escape reaction of the crabs, resulting in increased avian predation and subsequently in decreased fiddler crab population. Over one season, the *Uca* population decrease due to temephos treatments (as commonly used for mosquito control on New Jersey marshes) was about 20%. Longer term population changes may or may not be larger. They depend greatly on a number of population dynamics relationships that would need investigation in order to make any predictions. A first model of the population effects of insecticide treatments is used in simplified form as an illustration in Chapter 3.

Studies on organisms inhabiting salt marsh ponds or "potholes" included investigation of insecticide effects on aquatic nontarget insects, already discussed above, on the shrimp *Palaemonetes* sp., and on the killifish *Fundulus heteroclitus*. Neither temephos nor chlorpyrifos treatments had significant effects on species richness, species diversity, or density of the dominant, subdominant, and predatory insects inhabiting salt marsh ponds (Campbell and Denno, 1976). Experiments on the shrimp were only pilot studies involving measuring effects of field treatments on shrimp held in cages in marsh potholes. In most cases, 10–20% mortality over that observed in control plots was recorded. However, on one occasion mortalities of 68 and 44% were observed, perhaps due to interaction with stress induced by the high salinity observed at the time. Further studies would be desirable to try to measure possible changes in uncaged populations of *Palaemonetes* and to study interactions of insecticide toxicity with salinity and other expected environmental stresses. Preliminary laboratory studies showed that temephos is toxic to the shrimp at relatively low dosages (100% mortality occurred in 48 hours at concentrations in the medium between 0.5 and 1 ppm), and that important interactions with salinity and temperature probably exist. Interest in *Palaemonetes* is based on its suspected importance as a food source for marsh fish and birds.

A number of effects of temephos and chlorpyrifos on the fish *Fundulus heteroclitus* and *Cyprinodon variegatus* were observed in field and laboratory experiments (State Mosquito Control Commission, 1975; Thirugnanum and Forgash, 1975). The effects included reductions in growth rate of young fish from that observed in control plots, behavioral effects (such as loss of equilibrium and slow reaction to disturbance) and mortality and reduction in *in vitro* and *in vivo* activity of brain acetylcholinesterase. The extent of the effects varied with the compound and formulation used and with other factors. In general, chlorpyrifos was far more toxic than temephos when both were used as recommended for mosquito control in the field. No significant mortality over controls was observed in temephos-treated plots. However, in the field experiments, the fish were held in polyethylene mesh cages, reducing or eliminating possible predation interactions with sublethal effects of temephos. *Fundulus heteroclitus* and *C. variegatus* exposed to 10 biweekly applications of granular temephos in the field showed no significant inhibition of brain acetylcholinesterase activity. Exposure of *F. heteroclitus* to four biweekly applications of temephos emulsion only showed significant brain enzyme inhibition and obvious behavioral changes on one occasion (possibly due to an uneven insecticide deposition on the previous application). Thus, it is possible that relatively small increases in temephos dosages due to field factors, such as ingestion of temephos-containing food items or uneven distribution of the insecticide, could effect significant sublethal and lethal changes in the fish. Clearly, these effects of temephos suggest that further study of field effects on the marsh fish is desirable, particularly study of population dynamics including predation on the fish and predation by the fish on mosquito larvae. Interactions of these relationships with insecticide effects are likely to be important.

The lack of any apparent effects of the insecticide treatments on the sparrow populations was already described in the discussion of delimitation of sampling areas. A number of other bird species may or may not have been affected by the treatments; uncovering such effects would require longer term and more intensive studies on the bird community than were possible in this program.

The chemical studies on temephos distribution and persistence showed that, in general, temephos does not persist or accumulate in the marsh or its organisms; however, a few statements modifying the general findings are in order. Although temephos disappeared within 4 months of the last application from the soil and grass roots of most treated areas, relatively high (1.6–2 ppm) overwintering residue levels were found in higher marsh areas (vegetated by *S. patens*) treated with the granular temephos formulation. Levels of temephos in algae showed no accumulation with additional treatments, but showed relatively high values at certain times. Both this effect and the overwintering persistence in high marsh areas were thought to be due to time- and temperature-dependent release of the active ingredient from the clay granules rather than to physical or biological accumulation. A similar effect had been noted in the *Melampus* temephos content described earlier. At any rate, the presence of temporarily high levels of temephos in these compartments should be considered when evaluating potential toxicity to susceptible marsh organisms. For example, organisms feeding on biota containing temporarily elevated insecticide concentrations are at these times receiving higher doses than expected on the basis of the amounts applied only.

E. Use of Results

The study described above was clearly an ecosystem-level, survey-type environmental impact investigation. However, it differed markedly from the majority of "survey" environmental impact studies in that it did not devote 3 years of time and considerable resources to describe and classify the ecosystem components and then only make guesses as to possible effects of the insecticide treatments on the "known" system. Instead, the major investment was in field experimental studies to measure changes induced by use of the insecticides and to indicate problems that should be pursued further and areas that seemed of lower priority given the survey experimental results. Selected laboratory experiments were used to test hypotheses suggested by field experiments and observations.

The results of the study can, therefore, be useful in two ways. First, the adverse effects of insecticide use uncovered by the study can be evaluated for decision making on further insecticide use. The use of temephos for mosquito larviciding on marshes does not change the basic habitat in that there seem to be no effects on the dominant plants and a number of other marsh components. However, marsh fiddler crab (*Uca*) population decreases and mortalities in shrimp (*Palaemonetes*) and occasional mortalities in killifish (*Fundulus*) can be expected when temephos is used as recommended for mosquito larviciding in New Jersey. Since fiddler crabs, shrimp, and killifish are all food organisms for marsh birds and perhaps larger fish, we can suspect (unknown) longer-term impacts on the predator species. Further, fiddler crabs are thought to be important detritivores, facilitating turnover and perhaps export of nutrients and organic matter produced by marsh plants; decreases in fiddler crab populations may thus result in a net reduced value of marsh habitat contribution to various food webs. The adverse effects noted in killifish populations may also represent a cost in reduction of an existing biological control agent for mosquito larvae.

In addition to the adverse effects just noted, major uncertainties were identified in our estimates of impact of temephos use on marshes. One was the effects of the insecticide on bacterial and other microorganismal activity on the marsh, and the other was the effects on bird populations (primarily possible long-term impacts).

Thus decisions concerning the use of temephos can be made by weighing these biological adverse effects and uncertain adverse effects as costs and risks, (in addition to the monetary costs of running the larviciding program) against the monetary, health, and social benefits resulting from the increment in mosquito control due to use of larvicide (in addition to ditching and dyking, killing adult mosquitoes, etc.). Unavoidably, some of these costs and benefits will involve value judgements. However, the study we have just described helped to describe biological costs involved much more specifically than previously possible and yet helped to provide fairly comprehensive scanning of the biological costs and risks involved. A decision might be made to continue use of temephos on marshes, pending further research, but to ameliorate the expected impacts by actively avoiding creek bank areas inhabited by fiddler crabs. Further, the Rutgers study would indicate that general broadcasting of temephos granules in high marsh areas, where temephos release rates may be quite low, is to be avoided, since it appears to promote some occurrence of higher concentration and persistence. Suitable provisional substitutes might be use of emulsion formulation or more restricted (perhaps manual)

ful for decision making on future developments of this sort. The devices used to promote the effectiveness of the study were using the comparative approach and focusing on gross ecosystem function characteristics rather than on a complete description of species present or other reductionist approaches. In addition, although seasonal variability was expected in both artificial and natural systems, it was considered useful to compare the two types of systems at the same point in time. The comparison would at least allow formulation of recommendations to alleviate problems present in the summer.

Follow-up studies (Daiber, 1974, 1975) uncovered seasonal changes in the stratification and community structure pattern, explored differences in productivity and in suitability as fish habitat among tidal creeks, open bays, and artificial lagoons, and thus allowed additional recommendations for design of coastal recreational developments.

V. EFFECTS OF SPRUCE BUDWORM CONTROL MEASURES

The northern forests of North America have had, for at least hundreds of years, periodic outbreaks of spruce budworm (*Choristoneura fumiferana*). The budworm is a forest tree defoliator. Extensive data on budworm populations and their relationship to forest and weather phenomena have been collected by Environment Canada for more than 30 years. Morris (1963) summarized these studies and analyzed possible factors regulating budworm populations. In the province of New Brunswick, the dominant tree species are balsam fir, spruce, and birch. In spite of its common name, the spruce budworm defoliates balsam fir preferentially, and spruce to a lesser extent. Birch is not attacked by the budworm. In the absence of budworm, balsam outcompetes spruce and birch. Thus, between budworm outbreaks balsam tends to have the advantage, and during budworm outbreaks spruce and birch are favored.

The history of budworm outbreaks was studied by tree ring analysis. Since 1770 four outbreaks were detected; each lasted 7 to 17 years and was separated from the next outbreak by 34 to 72 years. Between outbreaks, the budworm is present only in very low densities but is not extinct. It is kept at low densities by mortality factors, such as parasites and insectivorous birds. The factors responsible for the sudden population explosions (four orders of magnitude increase in 3–4 years) appear to be an adequate food supply (forest recovered from previous outbreak) coupled with several successive years of warm dry summers. These weather conditions favor very rapid larval development, allowing the budworm to attain sufficient numbers to escape population control by predators and parasites and produce a budworm outbreak. After a period of mortality of the infested trees, the outbreak populations collapse. The outbreaks and collapses spread geographically through time due to the dispersal ability of the adult budworm moths. Thus, at any point in time the forests vary in age and species composition of stands from place to place.

Control measures to reduce loss of timber have been primarily insecticide applications to kill larval and adult budworm, although a number of other alternatives have been discussed [parasites and viruses (Watt, 1968) and forest harvesting practices (Stander, 1973)]. The overall population effects of insecticide spraying for budworm control are unclear, and some have suggested that spraying control programs may have resulted in more frequent and longer outbreaks than if the system had been left undisturbed (Baskerville, cited in Stander, 1973). In view of the cost of spraying programs and the possible undesirable impacts of insecticides on nontarget species, it is highly desirable, to put it mildly, to be reasonably sure that the control measure is having the desired effect on the target species at the population level.

In 1972, the Canadian Forest Service and a team of modeling ecologists from the University of British Columbia did a short-term (a few days) study of the effects of control practices on the spruce budworm–forest system. The study consisted of constructing a computer-based simulation model of the spruce budworm–forest system in New Brunswick as a device to test the outcome of various management programs insofar as possible and, importantly, to clarify, communicate, and organize available information and expertise and to identify critical research questions for future work. The details of the model are described by Stander (1973), and more general discussions of the study and related follow-up studies are presented by Walters and Peterman (1974) and Holling (1978).

The simulation model consisted of four submodels. One submodel simulated local budworm population dynamics in 265 different stands as a function of weather, the condition of the forest stand, and input of dispersing moths. Depending on the state of variables in the first submodel, the second submodel calculated adult dispersal between 265 local areas and the egg deposition in each area. The third submodel calculated the growth and species composition to be expected in the forest as a function of the budworm population levels resulting from the first two submodels. The fourth submodel simulated changes caused in the other three submodels by management actions such as altering forest condition or reducing budworm densities at specific times by insecticide spraying.

Many simplifications were made in constructing the model; these aimed at capturing the essential behavior of the interacting components while keeping the number of assumptions within manageable bounds. For example, a detailed population dynamics model for the budworm was not used; instead data from Morris (1963) and a few assumptions were used to construct functional relationships between initial egg density and resulting third instar densities and surviving adults (Walters and Peterman, 1974).

The results of the simulations (see Fig. 6.2) showed an outbreak occurring after about 40 years if no control methods were used; this time period is close to those observed historically in New Brunswick. Also, discontinuing spraying would result in collapse of the present outbreak in about 15 years beginning in the center of the province. In 1972 and 1973, this effect in fact began to be noticed in the central areas. Continuing present spraying programs resulted in very similar results to immediate cessation of spraying. When spraying similarly to the current pattern but using a very high dose of insecticide was simulated, the model predicted that the budworm outbreak would be prolonged indefinitely as a result of maintaining the balsam fir stands in mature condition. These results can only be considered as initial attempts, but begin the process needed to eventually achieve useful predictive ability.

Further benefits of the modeling exercise included identification of three research areas that should be pursued to fill in key gaps in model construction. These were female dispersal distances and resulting oviposition densities, effect of tree stand characteristics on oviposition by females, and population dynamics of budworm during low population conditions between outbreaks. Further, entomologists and foresters alike were able to appreciate information needs of the other discipline much more realistically than before.

The conditions that faced this study are somewhat typical of a class of environmental impact problems. The system in question is economically important and has thus been the subject of field studies and monitoring, usually by a government agency, for some time. The

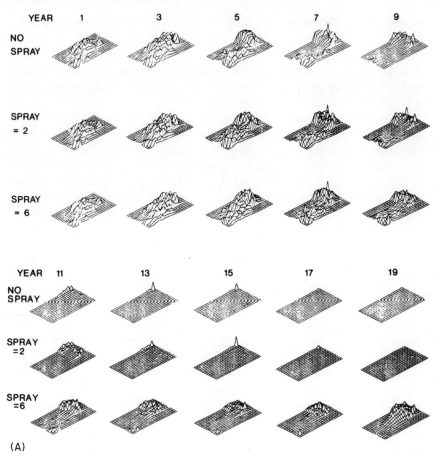

Fig. 6.2. Simulated spruce budworm egg densities on stylized maps of New Brunswick for alternative management strategies beginning in 1967. Each small grid area within each map represents 6×9 miles; map heights represent number of eggs per 10 ft² of foliage (maximum about 1000 eggs per 10 ft²). (A) shows egg densities resulting from simulating no spraying, spraying as actually practiced in 1972 (spray=2) and a very high spraying dose (spray=6), for 19 years. (B) shows simulated egg densities for 54 years in the no spraying condition. Note that further work on the model in later publications makes modified predictions and suggestions. (From Walters and Peterman, 1974.)

NO SPRAYING

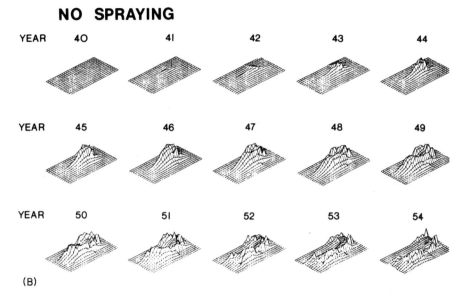

(B)

Fig. 6.2. *continued*

result is a body of historical and ecological data, much of which may be irrelevant but some of which provides knowledge of some major structural and functional features of the system. A variety of management practices has been employed, but the effects and impacts of these practices are neither clearly understood nor satisfactory to the managers or to the public. The management problem involves a complex interactive system, large geographical areas, and considerable variability in both space and time. Such situations often arise in management of forests, fisheries, wildlife, and other exploited natural resources. In this type of a situation, a modeling effort such as that employed on the spruce budworm can be an especially valuable contribution. It capitalizes on the available information, organizes and links selected relevant piecemeal insights and certainties with educated guesses on the uncertainties, provides a way of actually trying a number of alternative management actions on the simulated system, and reveals areas of research that should be emphasized to reduce substantial uncertainties about the effects of control measures. Walters and Peterman (1974) discuss further aspects of the rapid workshop approach to doing these analyses.

REFERENCES

Campbell, B. C., and Denno, R. F. (1976). *Environ. Entomol.* **5**, 477–483.

Daiber, F. C. (1972). "Environmental Impact of Dredge and Fill Operations in Tidal Wetlands upon Fisheries Biology in Delaware," Project F-13-R-15. Rep. Div. Fish Wildl., Dep. Nat. Resour. Environ. Control, Newark, Delaware.

Daiber, F. C. (1974). "Ecological Effects upon Estuaries Resulting from Lagoon Construction, Dredging, Filling, and Bulkheading," Project F-25-R. Rep. Div. Fish Wildl., Dep. Nat. Resour. Environ. Control, Newark, Delaware.

Daiber, F. C. (1975). "Ecological Effects upon Estuaries Resulting from Lagoon Construction, Dredging, Filling and Bulkheading," Project F-25-R-2. Rep. Div. Fish Wildl., Dep. Nat. Resour. Environ. Control, Newark, Delaware.

Fitzpatrick, G., and Sutherland, D. J. (1976). *Pestic. Monit. J.* **10**(1), 4–6.

Holling, C. S., ed. (1978). "Adaptive Environmental Assessment and Management." Wiley, New York.

Morris, R. F., ed. (1963). *Mem. Entomol. Soc. Can.* No. 31.

Oviatt, C. A., Nixon, S. W., and Garber, J. (1977). *Environ. Manage.* **1**, 201–211.

Stander, J. (1973). "A Simulation Model of the Spruce Budworm and the Forest in New Brunswick," IARE Mimeo Rep. Univ. of British Columbia, Vancouver.

State Mosquito Control Commission (1975). Nineteenth Annual Report, July 1, 1974–June 30, 1975. Trenton, New Jersey.

Thirugnanam, M., and Forgash, A. J. (1975). *Environ. Physiol. Biochem.* **5**, 451–459.

Walters, C. J., and Peterman, R. M. (1974). *Quaest. Entomol.* **10**, 177–186.

Ward, D. V., and Busch, D. A. (1976). *Oikos* **27**, 331–335.

Ward, D. V., and Howes, B. L. (1974). *Bull. Environ. Contam. Toxicol.* **12**, 694–697.

Ward, D. V., Howes, B. L., and Ludwig, D. F. (1976). *Mar. Biol.* **35**, 119–126.

Watt, K. E. F. (1968). "Ecology and Resource Management." McGraw-Hill, New York.

7

Conclusions

I. SEQUENCE OF ACTIVITIES IN BIOLOGICAL IMPACT STUDIES

The principal terms of reference for most biological impact studies are usually of the form "what will be the effects of factor or development X on biological components A, B, . . . , N? or "what will be the effects of factor or development X on ecosystem Z?" Most current impact studies respond to those terms of reference with the following sequence of activities. Literature and other investigators are surveyed to obtain descriptive information about the biological components or the ecosystem in question. Then field survey sampling is carried out to the extent allowed by the period of study specified. The survey sampling attempts to determine the distribution and abundance of the biological components A, B, . . . , N or the species present and the values of physicochemical factors present in ecosystem Z. Then descriptive summaries and tables are prepared from the field survey measurements, if any were made. In cases involving long study periods, the descriptive material may be voluminous. Last, a report is prepared in which speculations are presented as to the possible effects of development X or of several forms of development X on the described components or ecosystem. Unfortunately, the descriptions of the components or ecosystem by necessity lack information on causal linkages and pathways for indirect effects of the proposed developments, since the descriptions are based on the system in its present state.

A rather different sequence of steps is needed to utilize the approaches I have been proposing in this book. In the suggested sequence of steps that follows I will name and describe the activity, and refer to the chapters where detailed material on the activity may be found. Given terms of reference of the form "what will be the effects of factor or development X on biological components A, B, . . . , N or on ecosystem Z?" the following sequence may be profitable.

1. *Information Survey*

A rapid literature and/or field survey to describe the major ecosystem interactions of species A, B, . . . , N or the major functional characteristics of and dependencies within ecosystem Z. It should be noted that this differs markedly from a taxonomic catalog (Chapters 2 and 6).

2. *Modeling and Hypothesis Formulation*

From the information obtained in step 1, construct a qualitative explicit model of the major factors controlling the densities, distribution, and productivity of components A, B, . . . , N or of the major ecological processes occurring in ecosystem Z. The model may be at the conceptual, verbal descriptive level, may be a fully computerized simulation model, or may fall between these two categories. The model should include the proposed development factor(s) as one interacting component. This involves guessing at development effects, but in an explicit systematic framework and in the context of causal and functional pathways in the biological system. Based on this analysis, list predicted biological effects of the development factor(s) and identify processes and variables likely to be crucial to model outcome and to ecosystem function (Chapters 3 and 6 and Chapter 2, Section V).

3. *Testing Hypotheses*

Design formal hypothesis-testing field and/or laboratory experiments relating crucial processes and variables from step 2 to development factor(s) X. Experiments will test hypotheses rather than describe the system. Field experimental results and/or modeling will guide selection of laboratory experimentation, if any. Field experiments will use simulated or real development actions in some form (Chapters 4–6).

4. *Model Modification*

Based on the results of the experiments, modify the conceptual or simulation model of the interactions among major biological system components and development factor(s) X. Use this modified model to propose possible impacts of development factor(s) X on components A, B, . . . , N or on ecosystem Z (Chapters 3 and 6).

5. *Put the Predictions in Perspective*

Identify important assumptions of the analysis, key future data or analysis needs, and applicability limitations (Chapters 3 and 6).

It should be noted that the study can omit steps 3 and 4 (see example in Chapter 6, Section V) and still make several kinds of valuable contributions.

II. MEETING EXISTING CONSTRAINTS

The terms of reference for impact analysis studies usually ask questions we can never hope to answer fully even with unlimited time and resources. The accompanying time and resource constraints make it difficult to contribute enough insight and predictive power to facilitate decisions for short-term management. However, a number of devices can be used to perform the steps outlined in Section I under conditions of limited resources and time. First, the above approaches and steps can be taken, but the objectives can be miniaturized by studying and making predictions about only a limited time horizon or a limited geographical area. Examples are found in the study of the Delaware lagoon development effects: only summer effects were studied. Since this was the time when the most adverse effects were expected in this case, the limited time horizon was justifiable. Great care has to be used in these judgements to avoid overlooking time- and space-cumulative impacts. In this case, an advantage was availability of relatively old (20 years) stabilized affected systems for study so that cumulative effects may have occurred and thus be detectable.

A second device that can help meet study constraints is the comparative approach. In cases where the developments have already affected similar systems, much information may be available about impacts on those systems. If so the problem is really to try to determine the similarity of the system proposed for development to the postdevelopment, better-known systems.

A third approach, to omit steps 3 and 4 above, is that taken by the spruce budworm modeling study example. This approach is very valuable in meeting constraints, as it is rapid and cost-effective. Unfortunately, if a fully computerized simulation model is needed, it may be difficult to meet existing constraints of modeling expertise and facilities.

In any case, the approaches and steps suggested here do not necessitate more time and resources than a great many impact studies of the "busy taxonomist" type (see Chapter 6, Section I). In many cases, they also do not imply more investment and time than the "information broker" approach. This is clearly evidenced by the time horizons and resource uses of the examples described in Chapter 6.

Index

A
B
C 8
D 9
E 0
F 1
G 2
H 3
I 4
J 5